# Sampling in Statistics

Andale Publishing, LLC



ISBN 9798416512354

Cover image: Adobe Stock [Licensed]

Contents

# 1. INTRO TO SAMPLING

## Samples

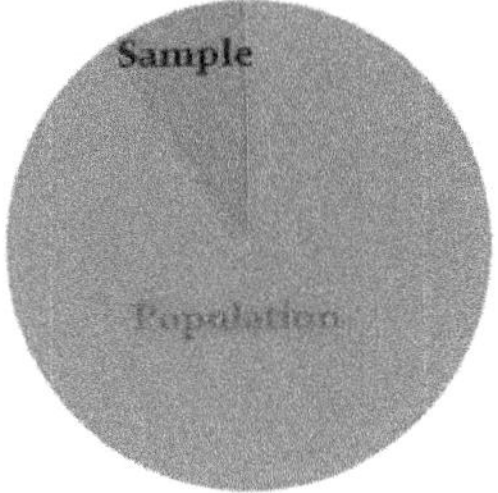


In statistics, you'll be working with samples. A sample is just a part of a population. If you want to find out how much the average American earns, you aren't going to want to survey *everyone* in the population (over 300 million people), so you would choose a small number of people in the population. For example, you might select 10,000 people.

Technically, you can't just choose *any* 10,000 people. For it to be statistical (i.e., one that you can use in statistics), the actual size must be found using a statistical method. Ten thousand people might not be the optimal amount for valid survey results: you may need more, or less. There are many, many ways to find sample sizes, including using data from prior experiments or using an online sample size calculator. How you find a sample size can be quite complex, depending on what you want to do with your data.

If you've decided to assemble your sample from scratch (for example, you aren't using prior data), then you need to choose a sampling method. Which sampling method you use depends on what resources and information you have available.

For example, the national draft worked by drawing random birth dates, a method called simple random sampling. For that to work, the government needed a list of every potential draftee's name and date of birth. The draft could also have used systematic sampling, drawing the $n$th name from a list (for example, every 100th name). For that to have worked, all the names must first have been compiled on a list.

### What is a "Sample Size"?

A *sample size* is a part of the population chosen for a survey or experiment. For example, you might take a survey of dog owner's brand preferences. You won't want to survey all the millions of dog owners in the country (either because it's too expensive or time consuming), so you take a sample size. That may be several thousand owners. The sample size is a representation of all dog owner's brand preferences. If you choose your sample wisely, it will be a good representation.

### When Error can Creep in

When you only survey a small sample of the population, uncertainty creeps into your statistics. If you can only survey a certain percentage of the true population, you can never be 100% sure that your statistics are a complete and accurate representation of the population. This uncertainty is called sampling error and is usually measured by a **confidence level**. A confidence level is the probability a parameter value falls within a specified range of values. Loosely speaking, it tells you how "confident" you are that your results will contain the true value for the population—even if you (or someone else) were to repeat your experiment. For example, you might state that your results are at a 90% confidence level. That means if you were to repeat your survey over and over, 90% of the time you would get the same results.

## Sampling Distribution

A **sampling distribution** is a graph of a statistic for your sample data. While, technically, you could choose any statistic to paint a picture, some common ones you'll come across are:

- Mean (the average)
- Mean absolute value of the deviation from the mean
- Range (a measure of spread)

- Standard deviation of the sample (standard deviation and variance are measures of how spread out data is from the mean)
- Unbiased estimate of variance
- Variance of the sample

Up until a certain point in statistics, you plot graphs for a set of numbers. For example, you might have graphed a data set and found it follows the shape of a normal distribution with a mean score of 100. Where probability distributions differ is that you aren't working with a single set of numbers; you're dealing with multiple statistics for multiple sets of numbers. If you find that concept hard to grasp: you aren't alone.

While most people can imagine what the graph of a set of numbers looks like, it's much more difficult to imagine what stacks of, say, averages look like.

AN EXPLANATION…

Let's start with a mean, like heights of students in the above cartoon. As you probably know, heights (and many other natural phenomenon) follow a bell curve shape. So, if you surveyed your class, you'd probably find a few short people, a few tall people, and most people would fall in between.

Let's say the average height was 5'9". Survey all the classes in your school and you'll probably get somewhere close to the average. If you had 10 classes of students, you might get 5'9", 5'8", 5'10", 5'9", 5'7", 5'9", 5'9", 5'10", 5'7", and 5'9". If you graph all those averages, you're probably going to get a graph that resembles the "sporkahedron." For other data sets, you might get a flatlined distribution, resembling a flat-roofed building.

It's almost impossible to predict what that graph will look like, but the **Central Limit Theorem** tells us that if you have a ton of data, it'll eventually look like a bell curve. That's the basic idea: you take your average (or another statistic, like the variance) and you plot those statistics on a graph.

The "mean of the sampling distribution of the means" is just math-speak for plotting a graph of averages (like I outlined above) and then finding the average of that set of data.

## Mean of the sampling distribution of the mean

In a nutshell, the **mean of the sampling distribution of the mean** is the same as the population mean (what you would expect to find as an average if you were to get data from the entire population). For example, if your population mean ($\mu$) is 99, then the mean of the sampling distribution of the mean, $\mu_m$, is also 99 (if you have a sufficiently large sample size).

## The Central Limit Theorem.

Roughly stated, the central limit theorem tells us that if we have many independent, identically distributed variables, the distribution will approximately follow a bell shape. It doesn't matter what the underlying distribution is.

Here's a simple example of the theory: when you roll a single die, your odds of getting any number (1, 2, 3, 4, 5, or 6) are the same (1/6). The mean for any roll is (1 + 2 + 3 + 4 + 5 + 6) / 6 = 3.5. The results from a one-die roll are shown in the first figure below: it looks like a uniform distribution. However, as the sample size is increased (two dice, three dice…), the distribution of the mean looks more and more like a normal distribution. That is what the central limit theorem predicts.

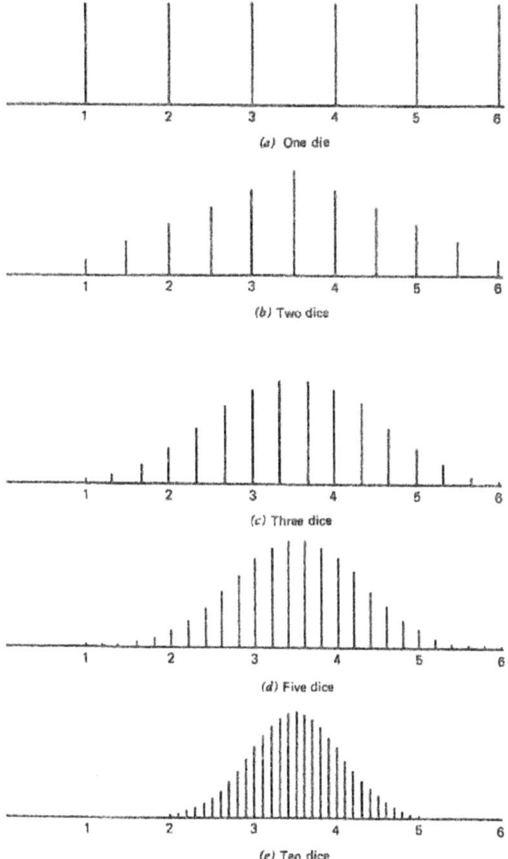


*Image: U of Michigan.*

As the sample size increases, distribution of the mean will approach the population mean of $\mu$, and the variance will approach $\sigma^2/N$, where N is the sample size.

You can think of a sampling distribution as a relative frequency distribution with many samples.

## Formula

The formula is $\mu_M = \mu$, where $\mu_M$ is the mean of the sampling distribution of the mean.

## Mean of Sampling Distribution of the Proportion

The **mean of sampling distribution of the proportion**, P, is a special case of the sampling distribution of the mean. The mean of the sampling distribution of the proportion is related to the binomial distribution, which shows how **binomial data** (data that takes on two values like yes/no or success/failure) is distributed.

Proportions are something you probably already know. For example: 100 people are asked if they are democrat. If 50 people respond "yes" then the sample proportion p = 50/100. Technically: A sample proportion is where a random sample of objects n is taken from a population P; if *x* objects have a certain characteristic, then the sample proportion "p" is *x*/n.

The sampling distribution of a proportion is when you repeat your survey or poll for all possible samples of the population. For example: instead of polling asking 1000 cat owners what cat food their pet prefers, you could repeat your poll many, many times.

## Mean of Sampling Distribution of the Proportion

If a random sample of *n* observations is taken from a binomial population with parameter p, the sampling distribution (i.e., all possible samples taken from the population) will have a mean $u_p = p$. With a large sample, the sampling distribution will have an approximate bell-shaped normal distribution.

## Standard Deviation of Sampling Distribution of the Proportion

The standard deviation of sampling distribution of the proportion, P, is also closely related to the binomial distribution and is a special case of a sampling distribution.

Example: You hold a survey about college student's GRE scores and calculate that the standard deviation is 1. It is highly unlikely that you will get the same results if you repeat the survey (you might get 1.1 ,1.2 or 0.9). Therefore, you'll want to repeat the poll the maximum number of times possible (i.e., you draw all possible samples of size *i* from the population). You'll have a range of standard deviations — one for each sample.

## Sampling Distribution of a Proportion

This is when you repeat your survey for all possible samples of the population. For example: instead of polling 100 people once to ask if they are democrat, you'll poll them multiple times to get a better estimate of your statistic.

## Standard Deviation

If a random sample of *n* observations is taken from a binomial population with parameter p, the sampling distribution (i.e., all possible samples taken from the population) will have a standard deviation of:

Standard deviation of binomial distribution =

$\sigma_p = \sqrt{[pq/n]}$

where q=1-p.

When the sample is large, the distribution will have an approximate normal distribution.

## Sampling Distribution of the Sample Proportion

The **Sampling Distribution of the Population Proportion** gives you information about the population proportion, *p*. For example, you might want to know the proportion of the population (*p*) who use Facebook. You can't survey everyone on the planet, so you use a sample and get the sample proportion and use that as an estimator.

When studying the sampling distribution of the sample proportion, you'll also see a lowercase $\bar{p}$. The lowercase version refers to a single value (i.e., a single estimate).

### USEFUL FORMULAS FOR SAMPLING DISTRIBUTION OF THE SAMPLE PROPORTION

Expected value of the sampling distribution of $\bar{P}$: $E(\bar{p}) = p$.

Variance for the sampling distribution of $\bar{P}$: $p(1-p) / n$.

Standard Error (SE) of the Sample Proportion: $\sqrt{(p(1-p) / n)}$. Note: as the sample size increases, the standard error decreases.

You can use the normal distribution if the following two formulas are true:

1. $np \geq 5$
2. $n(1-p) \geq 5$.

Z Score for sample proportion: $z = (\bar{P} - p) / SE$

### SAMPLE PROPORTION AND THE CENTRAL LIMIT THEOREM

In most statistics books (and in real life), you'll only make inferences about population proportions if you have a **large enough sample size**. If you have a large enough sample size, you can use the normal distribution for the sampling distribution of P.

How large is "large enough"? Use these formulas for a general guideline:

1. $np \geq 5$
2. $n(1-p) \geq 5$.

For example, if you had a sample size (n) of 50 and a proportion of 30%, then:

$n * p = 50 * .3 = 15$

$50(1-.3) = 50(.7) = 35$.

These are both larger than 5, so you can use the normal distribution.

You can transform $\bar{P}$ into a **z-score*** with the following formula:

Z-Score for sample proportion: z = ($\bar{p}$ – p) / SE.

*Z-scores tell you how far from the mean data points are.

*Example Question:*

A certain company's customer base is made up of 43% women and 57% men. An aggressive marketing campaign results in an increase of women customers to 46%, according to a sample survey of 50 customers. If the company hadn't run the campaign, how likely is it that 46% of customers are women? Was the campaign worth it?

Note that you're looking for the probability that $\bar{P}$ is greater than or equal to 46%.

**Step 1:** n * p = 50 * .43 = 21.5
50(1-.43) = 28.5.
Both are above 5, so we can use the normal distribution.

**Step 2:** Find the standard error (SE):
√ (p(1-p) / n) = √ (0.43(1-0.43) / 50) = 0.07.

**Step 3:** Find the z-score, using the SE you calculated in Step 2:
z = ($\bar{P}$ – p) / SE
P(Z≥) (0.46 – 0.43)/0.07 = 0.43.

**Step 4:** Look up 0.43 in the **z-table**.
(https://www.statisticshowto.com/tables/z-table/)

The probability is 0.3336, or 33.36%.*

At a probability of 33.36%, it's likely that the proportion of women would have been 46% without a campaign. It's unlikely that the marketing campaign made much of a difference.

*How I calculated this:

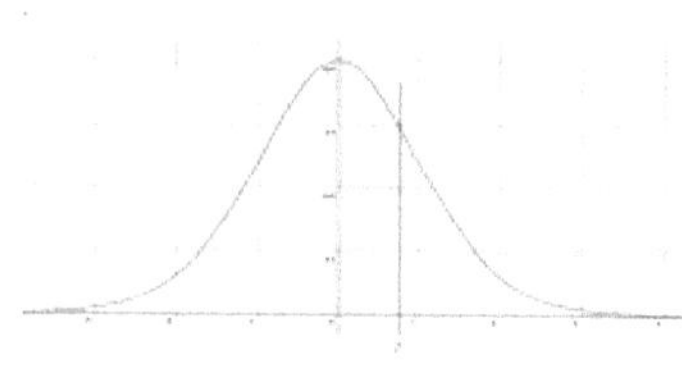

| Z | 0.00 | 0.01 | 0.02 | 0.03 | 0.04 |
|---|---|---|---|---|---|
| 0.0 | 0.0000 | 0.0040 | 0.0080 | **0.0120** | 0.01 |
| 0.1 | 0.0398 | 0.0438 | 0.0478 | **0.0517** | 0.05 |
| 0.2 | 0.0793 | 0.0832 | 0.0871 | **0.0910** | 0.09 |
| 0.3 | 0.1179 | 0.1217 | 0.1255 | **0.1293** | 0.13 |
| 0.4 | 0.1554 | 0.1591 | 0.1628 | **0.1664** | 0.17 |

1. I found a value of 0.1664 in the z-table. This is the area between the mean and z (shaded area in the center of the bell curve).
2. As I want the tail area (greater than z), I subtracted from 50% (i.e., half the curve). If I had been using a full z-table (from -3 to 3 standard deviations) I would have subtracted from 1.

## Sampling Design

**Sampling design** is a mathematical function that gives you the probability of any given sample being drawn.

Since sampling is the foundation of nearly every research project, the study of sampling design is a crucial part of statistics and is often a one or two semester course. It involves not only learning how to derive the probability functions which describe a given sampling method but also understanding how to design a best-fit sampling method for a real-life situation.

### EXAMPLES OF SAMPLING DESIGN

Sampling design can be very simple or very complex. In the simplest, one stage sample design where there is no explicit stratification and a member of the population is chosen at random, each unit has the probability

$n/N$

of being in the sample, where:

- n is the total number of units to be sampled,
- N is number of units in the total population.

Other types of design include:

- **Systematic sample**: all members of a population are listed in order and samples are chosen at defined intervals.
- **Stratified sample**: the population is first divided into strata and then samples are randomly selected from the strata (for example, divide a population between men and women, then randomly select a given number of men and a given number of women).
- **Cluster strata**: a population is divided into clusters and first clusters are randomly selected, then random members of the selected clusters are sampled. For instance, first randomly select a number of classes, then, from the class lists of those classes, randomly sample a number of students.

Each of these have their own sampling design function. The sampling method chosen will depend on the situation and priorities of the researcher. Sometimes, non-probability sampling methods will be chosen; for example, **convenience sampling**, where the sample is simply those easily reached and observed. Unlike systematic, stratified, or cluster sampling, these types of sampling cannot be easily described by a function.

REFERENCES

Mohadjer, Krenzke, & Van de Kerckhove. Technical Report. Chapter 4. Survey of Adult Skills (PIAAC), OECD.
Retrieved from
http://www.oecd.org/skills/piaac/Technical%20Report_Part%204.pdf on September 3, 2018
Maastricht University. Course Catalogue. Retrieved from https://www.maastrichtuniversity.nl/meta/325263/samplingdesign on September 3, 2018.
Raymo, Jim. Sample Design. Sociology 357 Class Notes.

Retrieved from https://www.ssc.wisc.edu/~jraymo/links/soc357/class8_F09.pdf on September 3, 2018.

## Sampling Unit

A **sampling unit** is the building block of a data set, an individual member of the population, a cluster of members, or some other predefined unit. It must be concretely defined as part of the groundwork for any statistical research or study. Typically, it is the minimum unit of observation that possesses the properties being studied. A lot depends on who your target group is and what data you have about the population. For example:

- In surveys and market research, the units might be households or targeted individuals (e.g., children under 18, adults over 60).
- In quality analysis, the unit might be a single production unit—a single food processor in factory that made these, or a single loaf of bread in a bakery.

The type of sampling you use will also define your unit. For example, in cluster sampling the cluster is the unit; in stratified sampling, the units are elements within each strata.

#### WHY IS DEFINING A SAMPLING UNIT IMPORTANT?

One of the fundamental reasons statistics exists in the first place is to compare different sets of data (for example, to see which is "best"). Loosely defined units make it impossible to make comparisons. For example, **relative precision** is the ratio of the error variances of two different sample designs which have the same sampling unit and sample size. If the units weren't defined well, or defined with a small error, the ratio would be meaningless.

#### REFERENCES

Daniel, J. (2011). Sampling Essentials: Practical Guidelines for Making Sampling Choices. SAGE.

London School of Hygiene and Tropical Medicine. The use of epidemiological tools in conflict-affected populations: open-access educational resources for policy-makers. Retrieved March 2, 2019 from:
http://conict.lshtm.ac.uk/page_30.htm

---

# Sampling Variability

**Sampling variability** is how much an estimate varies between samples. "Variability" is another name for *range*; Variability between samples indicates the range of values differs between samples.

Sampling variability is often written in terms of a statistic. The **variance**($\sigma^2$) and **standard deviation** ($\sigma$) are common measures of variability. You might also see reference to the variability of the sample mean ($\mu$), which is just another way of saying the sample mean differs from sample to sample. Sampling variability only refers to a statistic (i.e., a number generated from a sample)—never a population.

### VARIABILITY AND SAMPLING ERROR

A closely related term (almost a synonym) is **sampling error**. An error in sampling isn't a mistake — it's a measure of how much a value differs from the "true" value. Let's say the true weight of a population is 150 lbs. You take a sample and find the mean weight for the sample is 151 lbs. The 1 lb. difference is an "error." If you sample again, you might get different mean weights of 148 lbs., or 150.5 lbs., or 153 lbs. The different errors — 1/2 lb., 1 lb., 2 lbs., 3 lbs. — reflect the variability between your samples, or sampling variability.

### VARIABILITY AND SAMPLE SIZES

The "perfect" sample size is practically impossible to find.

Increasing or decreasing sample sizes leads to changes in the variability of samples. For example, a sample size of 10 people taken from the same population of 1,000 will very likely give you a very different result than a sample size of 100.

There is no "perfect" sample size that will give you accurate estimates for the sample mean, variance, and other statistics. Instead, you take your best "guess" — using standardized statistical procedures. In general, estimates will change from sample to sample and will probably never exactly match the population parameter.

### REFERENCES

Lodico, G. et al. (2010). Methods in Educational Research: From Theory to Practice.

## How to Find a Sample Size

You can use the data from a sample to make inferences (predictions) about a population. For example, the **standard deviation** of a sample (a measure of how spread-out sample data is) can be used to approximate the standard deviation of a population. Finding a sample size can be one of the most challenging tasks in statistics and depends upon many factors including the size of your original population.

### General Tips

1. *Conduct a census* if you have a small population. A **census** includes everyone in the population. A "small" population will depend on your budget and time constraints. For example, it may take a day to take a census of a student body at a small private university of 1,000 students but you may not have the time to survey 10,000 students at a large state university.

2. *Use a sample size from a similar study.* Chances are that your type of study has already been undertaken by someone else. You'll need access to academic databases to search for a study (usually your school or college will have access). A pitfall: you'll be relying on someone else correctly calculating the sample size. Any errors they have made in their calculations will transfer over to your study.

3. *Use a table to find your sample size.* If you have a generic study, then there is probably a table for it. For example, if you have a clinical study, you may be able to use a table published in Machin et al.'s Sample Size Tables for Clinical Studies, Third Edition.

4. *Use a sample size calculator.* Various calculators are available online, some simple, some more complex and specialized.

5. *Use a formula.* There are many different formulas you can use, depending on what you know (or don't know) about your population. If you know some parameters about your population (like a known standard deviation), you can use the techniques in the next section. If you don't know much about your population, use Slovin's formula.

## Cochran's Formula

The **Cochran formula** allows you to calculate an ideal sample size given a desired level of **precision** (which refers to how close different estimates from different samples are to each other), desired confidence level, and the estimated proportion of the attribute present in the population.

Cochran's formula is considered especially appropriate in situations with large populations. A sample of any given size provides more information about a smaller population than a larger one, so there's a 'correction' through which the number given by Cochran's formula can be reduced if the whole population is relatively small.

The Cochran formula is:

$$n_0 = \frac{Z^2pq}{e^2}$$

Where:

- *e* is the desired level of precision (the "margin of error"),
- *p* is the (estimated) proportion of the population which has the attribute in question,
- q is 1 – p.
- The **z-score** is found in a **Z table.** You can find a z-table, along with instructional videos on how to use one, at https://www.statisticshowto.com/tables/z-table/).

*A z-score gives you an idea of how far from the mean a data point is. More technically, it's a measure of how many standard deviations below or above the population mean a raw score is.*

Suppose we are performing a study on the inhabitants of a large town and want to find out how many households serve breakfast in the mornings. We don't have much information on the subject to begin with, so we're going to assume that half of the families serve breakfast: this gives us maximum variability. So, our proportion *p* is 50%: p = 0.5. Now let's say we want 95% confidence, and at least 5 percent—plus or minus—precision. A 95% confidence level gives us Z values of 1.96, per the z-table, so we get

$((1.96)^2 (0.5) (0.5)) / (0.05)^2 = 385.$

A random sample of 385 households in our target population should be enough to give us the confidence levels we need.

### MODIFICATION FOR SMALLER POPULATIONS

If the population we're studying is small, we can modify the sample size we calculated in the above formula by using this equation:

$$n = \frac{n_0}{1 + \frac{(n_0 - 1)}{N}}$$

Here, $n_0$ is Cochran's sample size recommendation, N is the population size, and $n$ is the new, adjusted sample size. In our earlier example, if there were just 1000 households in the target population, we would calculate

385 / (1 + ( 384 / 1000 )) = 278

For this smaller population, all we need are 278 households in our sample: a substantially smaller sample size.

## Slovin's Formula

If you take a population sample, you must use a formula to figure out what sample size you need to take. Sometimes you know something about a population, which can help you determine a sample size. For example, it's well known that IQ scores follow a normal distribution pattern. But what about if you know nothing about your population at all? That's when you can use **Slovin's formula** to find out what sample size you need to take, which is written as

$n = N / (1 + Ne^2)$

Where:

n = Number of samples,
N = Total population and
e = Error tolerance (alpha level).

The error tolerance, e, can be given to you (for example, in a question). If you're a researcher, you might want to figure out your own error tolerance; Just subtract your confidence level from 1. For example, if you wanted to be 98 percent confident that your data was going to be reflective of the entire population then:

- – 0.98 = 0.02.
- e = 0.02.

Example question: Use Slovin's formula to find out what sample of a population of 1,000 people you need to take for a survey on their soda preferences.

Step 1: Figure out what you want your confidence level to be. For example, you might want a confidence level of 95 percent (giving you what's called an *alpha level* of 0.05), or you might need better accuracy at the 98 percent confidence level (alpha level of 0.02).

Step 2. Plug your data into the formula. In this example, we'll use a 95 percent confidence level with a population size of 1,000.

- $n = N / (1 + N e^2) =$
- $1{,}000 / (1 + 1000 * 0.05)^2 = 285.714286$

Step 3: Round your answer to a whole number

(because you can't sample a fraction of a person or thing!)

285.714286 = 286

## PROBLEMS WITH SLOVIN'S FORMULA

Slovin's formula gives you a ballpark figure to work with. However, this non-parametric formula lacks mathematical rigor (Ryan, 2013). For example, there is no way to calculate **statistical power** (which tells you how likely your study distinguishes an actual effect from one of chance). It's unclear from any reference texts exactly what the "error tolerance" is (a mean, or perhaps a proportion?).

Some texts call the error tolerance a "tolerance margin of error" (e.g., Ariola, 2006), although it seems to be unrelated to the **margin of error** used in traditional hypothesis tests. The Margin of Error in that sense is the error associated with a result (for example, you could say 62% of people voted for so and so with a 3% margin of error). From the context, it's almost certainly another name for the **alpha level.**

The lack of precision with wording is yet another reason the formula has a poor reputation in academia. But perhaps the biggest reason that the formula isn't widely accepted is that is seems to have materialized out of nowhere. In fact, no one seems to even know who Slovin is, or even if he existed at all.

### Yamane's Formula

**Taro Yamane** is often credited with an identical formula as Slovin. However, his formula was published several years after Slovin's (in 1967).

$$n = \frac{N}{1 + N(e)^2}$$

Where:

- e = precision level.
- N = population size.

## Probability and Non-Probability Sampling

Sampling takes on two forms in statistics: probability sampling and non-probability sampling:

- **Probability sampling** uses random sampling techniques to create a sample. For each element in the sample, the probability is known and non-zero. In principle, every element of the population has the same chance at being included in the sample.
- **Non-probability sampling** techniques use non-random processes like judgment of the researcher or convenience sampling (where you grab participants who are conveniently located…usually near to you). The probability of being selected for the sample is unknown.

Probability sampling is based on the fact that every member of a population has a known and equal chance of being selected. For example, if you had a population of 100 people, each person would have odds of 1 out of 100 of being chosen. With non-probability sampling, those odds are not equal. For example, a person might have a better chance of being chosen if they live close to the researcher or have access to a computer. Probability sampling gives you the best chance to create a sample that is truly representative of the population.

As a rule of thumb, your sample size should be over about 30. If you have a small sample, you may need to try one of the non-probability sampling techniques instead.

The probabilities do not have to be equal for a method to be considered probability sampling. For example, one person could have a 10% chance of being selected and another person could have a 50% chance of being selected. It's nonprobability sampling when you can't calculate the probabilities at all.

REFERENCES

Wisniowski, A. et al. Integrating Probability and Nonprobability Samples for Survey Inference. Journal of Survey

Statistics and Methodology, Volume 8, Issue 1, February 2020, Pages 120–147, https://doi.org/10.1093/jssam/smz051

# 2. COMMON SAMPLING TYPES

# Area Sampling

Area sampling involves sampling from a map, an aerial photograph, or a similar area frame. It is often the sampling method of choice when a **sampling frame** (a list of a population) isn't available. For example, a city map can be divided into equal size blocks, from which random samples can be drawn.

Although area sampling is most often associated with maps, sometimes the samples might be drawn from lists (Särndal & Swensson, 2003).

## Clusters and Subsampling

The samples drawn from an area frame are often referred to as **clusters**. These clusters may be subsampled several more times.

For example, let's say you wanted to sample from a population of middle school students. The first sample might be drawn from a list of school districts, the second sample from a list of schools, the third a list of classes and then finally a list of students within those classes. The "frame" in this example is the four successive layers.

## Advantages and Disadvantages

Although area sampling using area frames is often the method of last resort, it does have a few distinct advantages:

- Area frames can be used for multiple variables at the same time. For example, an area sample on a city can collect data on land use, population, and income statistics.
- There's no overlap between sampling units; Every unit has an equal chance of being selected. This complete coverage results in unbiased estimates.

Disadvantages include:

- Although the area frames can be used in subsequent surveys, they can quickly become outdated (for example, if a city undergoes tremendous growth).

• Area frames can be costly to build.

• **Outliers** (very low or very high values) can be a problem, especially if your map has a few particularly dense or sparse areas; for example, a city that has a national park in its boundaries might have zero population in some areas and a huge population in another.

## References

D'Amico, V. (2000). Marketing Research. Tata McGraw-Hill Education.
Davis, C. (2009). Area Frame Design for Agricultural Surveys. Retrieved November 5, 2019 from: http://citeseerx.ist.psu.edu/viewdoc/download?doi=10.1.1.400.6067&rep=rep1&type=pdf
Särndal, C. & Swensson, J. (2003). Model Assisted Survey Sampling. Springer Science & Business Media.

# Bernoulli Sampling

**Bernoulli sampling** is an equal probability, **sampling without replacement** sampling design; "without replacement" means that once an item is chosen for a sample, it is discarded and cannot be chosen again.

In this method, independent Bernoulli trials on population members determines which members become part of a sample. All members have an equal chance of being part of the sample. The sample sizes in Bernoulli sampling are not fixed, because each member is considered separately for the sample. The method was first introduced by statistician Leo Goodman in 1949, as "binomial sampling".

The sample size follows a **binomial distribution** and can take on any value between 0 and N (where N is the size of the sample). If $\pi$ is the probability of a member being chosen then the **expected value** (or mean value) for the sample size is $\pi N$. for example, let's say you had a sample size of 100 and the probability of choosing any one item is 0.1, then the EV would be 0.1 * 100 = 10. However, the sample could theoretically be anywhere from 0 to 100.

*Example of Bernoulli Sampling*: A researcher has a list of 1,000 candidates for a clinical trial. He wants to get an overview of the candidates and so decides to take a Bernoulli sample to narrow the field. For each candidate, he tosses a die: if it's a 1, the candidate goes into a pile for further analysis. If it's any other number, it goes into another pile that isn't looked at. The EV for the sample size is 1/6 * 1,000 = 167.

An advantage to Bernoulli sampling is that it is one of the simplest types of sampling methods. One disadvantage is that it's not known how large the sample is at the outset.

## Cluster Sampling

**Cluster sampling** is used when natural groups are present in a population. The whole population is subdivided into **clusters**, or groups, and random samples are then collected from each group.

### When to Use Cluster Sampling

Cluster sampling is typically used in market research. It's used when a researcher can't get information about the whole population, but they *can* get information about the clusters. For example, a researcher may be interested in data about city taxes in Florida. The researcher would compile data from selected cities and compile them to get a picture about the state. The individual cities would be the clusters in this case. Cluster sampling is often more economical or more practical than stratified sampling or simple random sampling.

### Requirements

1. Cluster elements should be as heterogeneous (unsimilar) as possible. In other words, the population should contain distinct subpopulations of different types.
2. Each cluster should be a small representation of the entire population.
3. Each cluster should be **mutually exclusive**. Mutually exclusive events cannot happen together at the same time, so it should be impossible for each cluster to occur together. In the city tax example, it would be impossible for Miami city taxes and Jacksonville city taxes to occur together, so it fits the requirements for mutual exclusivity.

### Types of Cluster Sampling

- *Single-stage cluster sampling*: all the elements in each selected cluster are used.
- *Two-stage cluster sampling*: a random sampling technique is applied to the selected clusters. For example, once you've decided on your clusters, you could use simple random sampling (e.g., picking numbers out of a hat) to select your sample.

### ReferenceS

Lohr, S. (2019). Sampling: Design and Analysis (Chapman & Hall/CRC Texts in Statistical Science) 2nd Edition. Routledge.

## Convenience Sampling

**Convenience sampling** (also called accidental sampling or grab sampling) is where you include people who are easy to reach. For example, you could survey people from:

- Your workplace,

- Your school,
- A club you belong to.

Convenience sampling is a type of non-probability sampling, which *doesn't include random selection of participants.*

## Why Use Convenience Sampling?

Although convenience sampling is, like the name suggests—convenient—it runs a high risk that your sample will not represent the population. However, sometimes a convenience sample is the only way you can drum up participants. According to Barbara Sommer at UC Davis, it could be "…a matter of taking what you can get".

Convenience sampling does have its uses, especially when you need to conduct a study quickly or you are on a shoestring budget. It is also one of the only methods you can use when you can't get a list of all the members of a population. For example, let's say you were conducting a survey for a company who wanted to know what Big Box Store employees think of their wages. It's unlikely you'll be able to get a list of employees, so you may have to resort to standing outside of Big Box Store and grabbing whichever employees come out of the door (hence the name "grab sampling").

## Advantages of Convenience Sampling

- It's relatively easy to get a sample.
- It's inexpensive, compared to other methods.
- Participants are readily available.

## Disadvantages of Convenience Sampling

The method cuts out a large part of the population. As a result, this leads to several issues, including:

- An inability to generalize the results of the survey to the whole population.

- The possibility of under- or over-representation of the population. For example, if you grab people at a Big Box Store, some population segments, like people with extra cash who love to shop, are more likely to be overrepresented. Others, like people who are home-bound or people with lower incomes, may not make it into your survey at all.
- Biased results, due to the reasons why some people choose to take part, and some do not. "**Bias**" is another name for the type of errors that creep into your analysis and results.

## How to Analyse a Convenience Sample

Results from these samples are easy to analyze but hard to replicate. While you can use any analysis method you like, you won't be able to generalize your results to the larger population.

Perhaps the biggest problem with convenience sampling is dependence. **Dependence** means that the sample items are all connected to each other in some way. This dependency interferes with statistical analysis. Most formal hypothesis tests and have an underlying assumption of random selection, which you do not have. Perhaps most problematic is the fact that p-values produced for convenience samples can be very misleading. **P-values**, or probability values, are an indicator of statistical significance; you'll usually get one from running a hypothesis test using software. In many cases, you'll decide whether your hypothesis is meaningful based on this one small value, so these need to be accurate. Convenience sampling will *not* give you meaningful p-values.

## Recommendations for analysis

The biggest recommendation is simple: If possible, use probability sampling instead of convenience sampling (Berk & Freedman, 2003). Other recommendations:

- Take multiple samples over the course of your study. If you do this, you may be able to model the selection process, producing more reliable results.

• Don't use post-hoc tests (**post-hoc tests** are run after your main analysis) as a tool to adjust your results in an attempt to deal with dependent data.
• Repeat your study again, to see if your results are truly replicable (Freedman, 1991; Berk, 1991; Ehrenberg and Bound, 1993).
• For larger samples, use **cross validation** to model one half of the data. You can then compare the results with the second half of the data to see if they match.
• Don't meta analyze convenience samples. **Meta-analysis** combines the findings from existing research into one, comprehensive thesis. A meta-analysis can uncover trends or themes that weren't apparent in individual pieces of research. If you're using biased data from convenience samples, then any "trends" you uncover are going to be suspect. Summarize results instead.

**References:**

Berk R. A. (1991) "Toward a Methodology for Mere Mortals," in P. V. Marsden (ed.), Sociological Methodology, Volume 21, Washington, D. C.: The American Sociological Association.
Berk, R. and Freedman, D. (2003) Statistical Assumptions as Empirical Commitments. In TG Blomberg and S Cohen, eds. Law, Punishment, and Social Control: Essays in Honor of Sheldon Messinger. Aldine de Gruyter, New York. pp. 235-54.
Ehrenberg A. S. C. and Bound J. A. (1993) "Predictability and Prediction," Journal of the Royal Statistical Society, Series A, 156 Part 2: 167–206.
Freedman D. (1991) "Statistical Models and Shoe Leather," in P. V. Marsden (ed.). Sociological Methodology, Volume 21, Washington, D. C.: The American Sociological Association.
Sommer, B. (n.d.). Types of samples. Retrieved September 13, 2017 from:
http://psc.dss.ucdavis.edu/faculty_sites//sommerb/sommerdemo/sampling/types.htm

## Critical Case Sampling

**Critical cases** are the ones most likely to give you the information you need. You collect samples that are most likely to give you the information you're looking for; They are particularly important cases or ones that highlight vital information.

This type of sampling is "…particularly useful if a small number of cases can be sampled" (Strewig & Stead, 2001). That small number of cases (assuming they can be classified as "critical") are the ones more likely to provide a wealth of information.

### Why Use Critical Case Sampling?

As Patton (1990) states: "If it can happen there, it can happen anywhere." Let's say you were studying reading levels for a popular science magazine, with reading levels ranging from an 8th grade level all the way up to graduate school. If 8th graders can understand the science articles, then everyone about that level should also be able to understand the articles — making the 8th grade level readers a "critical case." Using critical cases also makes sense if you're on a tight budget and want to study the participants who are most likely to provide crucial information.

### Examples

Scientists deal with critical cases all the time. For example, Gregor Mendel discovered the fundamental laws of inheritance through his meticulous study of pea plants. If Mendel had attempted to focus on a broad range of species instead of just one, he may not have made the discoveries that he is renowned for, such as his finding that genes are inherited in pairs — one from each parent.

### References:

Cold Spring Harbor Laboratory. DNA Learning Center: Gregor Mendel. Retrieved 1/20/2017 from:

https://www.dnalc.org/view/16151-biography-1-gregor-mendel-1822-1884-.html
Patton, M. (1990). Qualitative evaluation and research methods. Beverly Hills, CA: Sage. PDF.
Strewig, F. & Stead, G. (2001) Planning, Reporting & Designing Research. Pearson, South Africa.

# Direct Sampling

**Direct sampling** is a somewhat informal term that can refer to when the sample is taken from the actual population and not, for example, related records like driver's licenses, voter registration cards, or census forms.

Although the concept sounds simple, direct sampling isn't always possible. It may be too expensive or time consuming to get a direct sample, or the sampling frame may simply not be available.

## As The Opposite of Inverse Sampling

Another definition for a direct sample is where individuals are randomly drawn from a population. The individuals aren't tagged and returned to the population. Sometime later, a random sample size $n$ is taken, and the number of tagged individuals is counted.

## Sampling from Distributions

Direct sampling can also refer to sampling from a particular probability distribution instead of a specified population. For example, in **Bayesian theory**, an alternative to the "traditional" frequentist statistics you've probably studied in high school or as an undergraduate, sometimes it's necessary to sample from a **posterior distribution**, which summarizes what you know after the data has been observed.

When you know enough about the posterior distribution, you can directly sample from it. However, in many cases you may not know enough about the distribution, or it may be impossible to sample from it.

### Non-Probabilistic Sampling

Some authors use the term direct sampling to refer to a particular non-probabilistic sampling method. For example, Bdour & Koury (2009) refer to direct sampling in auditing as where:

> *"Each item sampled is selected based on some judgmental decisions set forth by the auditor who excludes equal opportunity selections and depends more on the intentional selection of items based on criteria which would relate less or more with sample representativeness of the population."*

### References

Bdour, J. & Koury, A. (2009) Statistical sampling in auditing: the case of Jordanian auditor practices. In European Journal of Management. Retrieved December 13, 2017 from: http://www.freepatentsonline.com/article/European-JournalManagement/208535117.html

Dodge, Y. (2006) The Oxford Dictionary of Statistical Terms. Oxford University Press.

Li, X. & Xu, R. (2008) High-Dimensional Data Analysis in Cancer Research. Springer Science & Business Media.

Scheaffer et. al (2011). Elementary Survey Sampling. Cengage Learning.

## Distance Sampling

**Distance sampling** is a catch-all term to describe a range of methods to estimate the density of biological populations using measured distances to individuals in the population. It can also be used to estimate the abundance of inanimate objects like birds' nests or animal burrows. This type of sampling — assuming it is properly designed — can improve precision of estimates and reduce bias.

## Types of Distance Sampling

The two main methods are line transect and point transect, but other methods (mostly sub-types of either point or line transect) exist.

1. Line Transect Surveys
2. Cue-Counting
3. Point Transect Surveys
4. Strip Transects
5. Trapping webs

## Line Transect Surveys

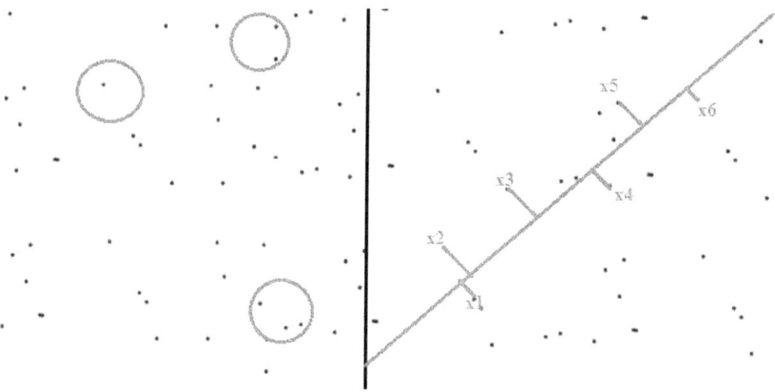


*Traditional sampling (left) and line transect (right).*

In traditional sampling, random areas (shown as circles in the above image) are chosen and the number of items are counted. These counts are averaged and used to estimate the population in the whole area. **Line transect** uses a set of lines; Distances to seen items are noted and used to estimate the population size. The lines are traversed by trained observers on foot, on horseback or on a vehicle. Depending upon the location, boats, scuba gear or aircraft are also used. For more detail on this method, see: *Line Intersect Sampling*, covered later in this chapter.

## Cue-Counting:

**Cue-counting** is an offshoot of line transect surveys developed for mapping marine life but extended to other animals. Instead of actual animal sighting, "cues" are counted instead. For example, the blow of a whale or a specific bird song.

## Point Transect Surveys:

**Point transect surveys** are like line transects. However, instead of a line there is a series of points. The observer usually stays at each point for a set time. For example, there may be 10 equidistant points along the route where the observer stays for 15 minutes at each point.

## Strip Transects:

**Strip transects** are where a strip of width $n$ is chosen, and a census is taken of items in that strip. Unlike line transects, distances are not measured. Strip transects have the rather impractical assumption that a count is taken of all items in the strip. It is therefore not used as often as line transects.

## Trapping Webs:

**Trapping webs** are a sub-type of line transect where traps are placed at intervals along a randomly chosen line. The observed distance in trapping webs is replaced by a known distance of the animal trap.

## References:

Buckland et. al. (2001). Introduction To Distance Sampling: Estimating Abundance of Biological Populations 1st Edition..

Stephen T. Buckland, E.A. Rexstad, Tiago A. Marques, C.S. Oedekoven. (2015). Distance Sampling: Methods and Applications.

# Double Sampling

**Double sampling** is a two-phase method of sampling for an experiment, research project, or inspection. An initial sampling run is followed by preliminary analysis, after which another sample is taken, and more analysis is run.

It is used in three main ways:

1. Acceptance/rejection double sampling,
2. Ratio double sampling, and
3. Stratification double sampling.

## Acceptance/rejection double sampling.

Used primarily in industrial sampling inspection, **acceptance/rejection double sampling** is the simplest type of two-phase sampling.

First, an initial random sample is taken and tested for variable *x*. If a certain high threshold is reached, the result is positive, and no further samples are run. If a certain low threshold is not met, the result is negative and, again, no further samples are run. But if the results are in some middle ground, a new, larger sample is taken and the results from both runs are used to draw a conclusion.

## Ratio double sampling

In **ratio double sampling**, an easy to study "auxiliary" variable *x* is used to gather information on the harder-to-research target variable *y*. Variable *x* is chosen as one which is highly correlated to *y*. **Correlation** is any statistical relationship (causal or not) between two sets of variables or data.

First, a large random sample is analyzed for *x*. Then, a subset of this sample is analyzed for *y*. The ratio of *x* to *y* over the subset is taken, and this allows us to reach conclusions as to what the value of y will be in the larger sample (and so in the population as a whole).

### Stratification double sampling.

A two-phase sampling method can also be used in situations involving **stratification**, where the population is divided into groups. In this way of sampling, an initial random sample allows us to target the second sample appropriately.

In our initial random sample, we look only at the stratification of a population, and the second sample, in which our target variable is studied, is taken over the stratification categories in a representative way.

For example, if you were studying how many books the average city dweller read in a week, you might stratify the population by college graduates and less educated people. If an initial quick survey told you that, in 100 residents, 60 were college graduates and 40 were not, you could decide to sample college graduates and non-graduates at a 6/4 ratio during phase two of your study.

### References

Shalabh, Sampling Theory Chapter 8. Retrieved from http://home.iitk.ac.in/~shalab/sampling/chapter8-sampling-doublesampling.pdf on January 28, 2018

Penn State STAT 506 Sampling Theory and Methods Course Materials Retrieved from https://onlinecourses.science.psu.edu/stat506/node/48 on January 28, 2018

Encyclopedia of Statistical Sciences: Double Sampling Retrieved from http://onlinelibrary.wiley.com/doi/10.1002/0471667196.ess0525.pub2/abstract on January 28, 2018

## Experience Sampling

The focus of **experience sampling** is examining experience in context, as people go about their daily lives.

Experience sampling is a way to find out more about an experience while the event is happening. Participants stop what they are doing and take time to note their experiences over a period of days, weeks, or even years — which can result in hundreds of data items per participant. Sometimes, more objective methods are used, such as activity monitors or random sound recordings (Zirkel, 2015). The method can help researchers understand people's thoughts, actions, and activities with minimal intrusion into participants daily lives.

Experience sampling has the following characteristics:

- Collects real-time, "in the moment" data.
- Collects data from the environment where the event is happening.
- Typically involves many observations.
- Usually involves a brief open- or closed-ended questionnaire.
- Is dependent upon careful data collection.

This type of sampling is also called:

- Ambulatory self-reporting,
- The daily diary method,
- Ecological momentary assessment,
- Intensive-longitudinal design.

## When do Participants Note Their Experiences?

Exactly when participants pause their activity is up to the researcher. The primary ways are event occurrence, preselected time intervals, or signaling.

Event occurrence involves key events — usually ones pertaining to the study in question. For example, a study on time management might ask someone to record details when they are late to an appointment, or a headache medication study might ask a participant to answer a short questionnaire when they get headaches.

- Estrin and Sim (2010) used the GPS feature on mobile phones to find out the time of day study participants left the house. The time was used as a measure of wellness for people with chronic illnesses and conditions like asthma, diabetes, and obesity.
- Zirkel et. al asked college students to complete surveys immediately after each class, where they indicated how much anxiety (or calmness), sadness (or happiness) they felt during the class. The authors intend to use the data to see how these experiences affected students' future decisions to take classes in a particular discipline or major.

Pre-selected time intervals are set at the beginning of the study. A person might be asked to report how they feel when they wake up, when they eat meals, or when they go to bed.

Signaling might involve texts, phone calls or a set timer. For example the Science in the Moment (SciMo) study distributed beepers to students in high school science classes. The goal was to investigate how male and female students felt about science education. The beepers went off randomly during class, signaling students to complete short surveys about their experiences.

## Advantages over Other Data Collection Methods

Experience sampling has several advantages over other methods:

1. You can get access to parts of people's lives that would otherwise be difficult or impossible to access (such as feelings during a workout or while participating in professional development classes).
2. **Recall bias** (where people "recall" and are affected by their previous experiences or answers) is reduced because you're collecting data as the event is happening.
3. Thoughts and feelings can be studied in very specific situations.
4. **Statistical power** (the probability that your analysis will find a significant difference) tends to be higher than for many other quantitative methods.

## Context-Aware Experience Sampling

Context-Aware Experience Sampling (CAES) uses technology like mobile phones to detect events, triggering a cue to collect data. For example, software can detect an increased heart rate, an activity like cycling, or movement into a certain geographic area (using GPS location). The CAES project at MIT is (as of time of writing) developing more tools in this area.

## References

Bolger, N. Intensive Longitudinal Methods: An Introduction to Diary and Experience Sampling Research (Methodology in the Social Sciences) 1st Edition. The Guilford Press. 2013.

Estrin, D., & Sim, I. (2010). Health care delivery: Open mhealth architecture. An engine for health care innovations. Science, 330, 759–760. Retrieved August 17, 2017 from: https://courses.cs.washington.edu/courses/cse599p1/16sp/readings/estrin2010.pdf

Hektner, J.M., Schmidt, J.A., & Csikszentmihalyi, M. (2007). Experience sampling method: Measuring the quality of everyday life. Thousand Oaks, CA: SAGE Publications.

Zirkel et. al (2015). Experience-Sampling Research Methods and Their Potential for Education Research. Retrieved August 17, 2017 from: http://indiana.edu/~mcmlab/Zirkel%20Garcia%20Murphy%20ER%202015.pdf

## Extreme Case Sampling

**Extreme Case Sampling** focuses on participants with unique or special characteristics. An extreme case (or deviant case) can be thought of as an outlier—an observation that takes on an extremely high or extremely low value. The general idea is that if you study extremes of the population, it could garner some valuable insights that can be generalized to the population as a whole. For example, if you were studying inner city violence, you could study a city with high violence and compare it to a city with low violence.

Like any sampling technique where a researcher deliberately chooses cases, extreme case sampling could result in **selection bias**, undermining results (Collier & Mahoney, 1996). Selection bias, where individuals chosen are different from those in the population, can largely be avoided if extreme cases from both ends of the spectrum are considered along with the general population.

### Examples

- Angela Browne studied male victims of domestic violence for her book "When Battered Women Kill."
- Kristen Monroe studied a variety of unusual altruists for her book "The Heart of Altruism." They included a poetry editor who — armed with a cane — saved a young girl from being raped, and Otto Schindler, the German businessman who saved over a hundred Jews.
- Benjamin Reilly's work on studying democracy and electoral systems focused on unusual societies like Papua New Guinea, a country of "exceptional ethnic fragmentation" which has thousands of ethnic micropolities" speaking 840 languages.
- Frederic Deyo (1987) studied East Asian NICs (Newly Industrialized Countries), which stood "in stark contrast" to other Third World countries.

### Quantifying Extreme Cases

Gerring (2006) defines extreme cases by their z-scores. However, this only works with **normally distributed** cases—those cases that fit the "bell curve".

### References

Deyo, F. (Ed.) (1987). The Political Economy of the New Asian Industrialism. Cornell University Press.
Gerring, J. (2006). Case Study Research: Principles and Practice. Cambridge University Press.
Monroe, K. (1996). The Heart of Altruism: Perceptions of a Common Humanity. Princeton University Press.
Reilly, B. (2000). Electoral Systems for Divided Societies. Retrieved 1/20/2017 from: (https://devpolicy.crawford.anu.edu.au/pdf/staff/ben_reilly/ReillyB_04.pdf.)

## Haphazard sampling

**Haphazard sampling** is where you try to create a random sample by *haphazardly choosing items* to try and recreate true randomness. It doesn't usually work, because of **selection bias**: the choices you make for your selection knowingly or unknowingly create unrepresentative samples. To create a true random selection, use one of the tried and testing random selection methods, like simple random sampling.

Even if you try to choose items without any bias or reason for including (or excluding) items, haphazard samples nearly always result in a sample that looks a lot different from a random sample. Therefore, the results from your test or experiment will have unpredictable errors and most likely, invalid results.

Sometimes haphazard sampling is used because it is cheaper than other sampling methods or because you can't meet random sampling requirements for technical reasons (like lack of access to computer software). If you must use haphazard sampling, you can raise the odds of a successful sample by:

1. Ensuring your sample selections are independent of each other. In other words, select one item and then randomly choose another. Don't choose two *A*s or two *B*s because it seems easier.
2. Ensure that each item has an equal probability of being selected. Person A should have the same odds of being selected as person B, C or Z.
3. Use larger sample sizes. Research has shown that increasing your sample size can reduce haphazard selection bias.

This sounds easy, but in practice your choices can be influenced by factors you aren't aware of. For example, you might unconsciously choose a name because of the way a name looks, or you might show a hidden gender preference. You might subconsciously exclude an item and include another because you know one item would be easier to locate than another. It's not possible to identify and eliminate all your biases, which is why random sampling is preferred.

**Reference:**

Hall, T., T. Herron, B. Pierce, and T. Witt. 2001. The effectiveness of increasing sample size to mitigate the influence of population characteristics in haphazard sampling. Auditing: A Journal of Practice & Theory 20 (1): 169–185

# Inverse Sampling

In **inverse sampling** (sometimes called standard inverse sampling), you continue to choose items until an event has occurred a specified number of times. It is often used when you don't know the exact size of the sample you want to take. For example, let's say you were conducting a wildlife management survey and wanted to capture 20 banded birds. You capture birds at random until you have collected 20 banded birds (and an unknown number of unbanded birds).

The resulting sample size could be 100 birds, or it could be 88 birds, or 203 birds and so on.

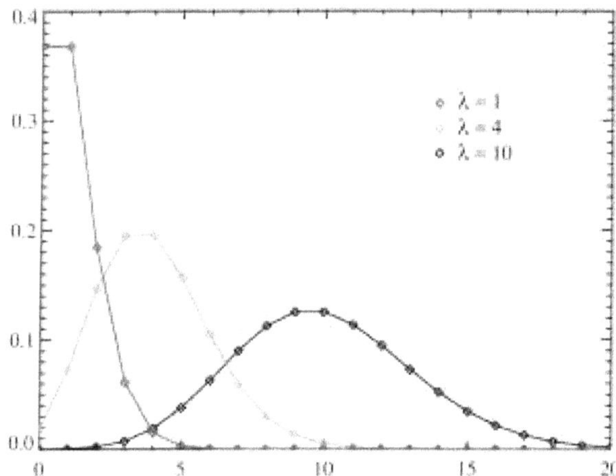


*The Poisson family of distributions can be used for inverse sampling.*

In a **binomial experiment**, where an event is either going to happen, or it's not going to happen, this type of sampling is called **negative binomial sampling.** The two terms (inverse sampling and negative binomial sampling) are often used interchangeably. However, Inverse Sampling can refer to sampling based on either the Negative Binomial Distribution (in which case it is called Negative Binomial Sampling) or the **Poisson Distribution**. If sampling is performed based on the Poisson distribution, you're interested in events occurring in a certain time period. For example, you might be interested in how many accidents happen on a particular road over the period of one week.

**Types and Designs**

Paul Lavrakas outlines two types of inverse sampling in the Encyclopedia of Survey Research Methods:

- **Vector-at-a-time sampling** involves drawing two observations at the same time (one from each of two populations). Sampling stops when x "Successes" are obtained from one of the populations.
- **Play-the-winner sampling** is where one item at a time is drawn from one of the (randomly selected) populations. Sampling stops when a "Failure" happens and then sampling continues in the second population until a Failure happens. At that point, sampling continues, switching back and forth between the two populations until a pre-specified number of Successes has occurred.

The two basic methods for Inverse Sampling are **Multiple Inverse Sampling (MIS)** and **General Inverse Sampling (GIS).**

- MIS (originally proposed by Chang and colleagues) is used when subpopulation sizes are known. Its main advantage is that it avoids the problem of empty strata in post-stratification. In stratification, the sample is selected, then split into strata (for example, by sex, age, or other characteristics). Some of the selected strata may end up with zero entries, especially if the category is rare. MIS ensures that at least *n* items will end up in each stratum (sampling is continued for a specified number of observations in each stratum).
- GIS avoids sampling an "infeasible number of units" (Salehi & Seber). Sampling is performed until strata contain a pre-specified number of units or until a maximum sample size has been reached. Applications

Inverse sampling is often performed when a certain characteristic is rare. For example, it is a good method for detecting differences between two different treatments for a rare disease; It avoids the problem of sparse data due to a disease's rarity. According to Lavkras, Play-the-winner sampling is preferred for clinical trials because the sample with the poorer population will always be smaller than the sample with the "better" population.

### Advantages and Disadvantages

In general, inverse sampling will give you more precise estimates than direct sampling (Scheaffer et. al, 2011), if the sample size n required to obtain *n* individuals is small compared to the population size N. However, as the sample size is unknown and could theoretically be infinite (in some cases), this technique can be costly, labor-intensive, and time consuming. Compared to random sampling estimated variances are usually much larger.

### References:

Paul J. Lavrakas. Encyclopedia of Survey Research Methods.
Kuang-Chao Chang et. al. MIS sampling in post stratification. PDF.
M Salehi & Seber G. A general inverse sampling scheme and its application to adaptive cluster sampling. Scheaffer et. al (2011). Elementary Survey Sampling. Cengage Learning.

## Kish Grid

A **kish grid** is a way of randomly choosing household survey respondents. The method avoids selection bias, which is usually a result of not using the correct procedures to choose your participants.

If you visit houses and survey the first person to answer the door, your results are probably going to be biased against the very young or very old, who are less likely to answer the door first. The Kish Grid addresses this problem by assigning numbers to each member of the household, based on age. The most important aspect of the grid is that it assigns an equal probability of selection for each possible survey participant (Lewis, Beck et. al, 2003).

### Advantages over First/Last Birthday

The first birthday/last birthday method chooses participants based on which household member has the first or last birthday of the year. This method may be problematic because:

- Some countries may not celebrate birthdays.
- Not everyone in a household knows everyone else's birthday. For example, a daughter-in-law might not know when grandpa-in-law was born, or no one may know the birth date of a new household member.
- As the grid doesn't use specific dates, these problems are avoided.

## How to Use a Kish Grid

The Kish Grid has a column for each household you visit and a row for the number of eligible people. Start labeling with the youngest members of the household to give them a slightly greater chance of being chosen. Young people tend to be more difficult to pin down at home, so if you don't stack the odds in their favor a little, your results might be biased in favor of older people.

| | Eligible People | | | | | | | |
|---|---|---|---|---|---|---|---|---|
| Household | 1 | 2 | 3 | 4 | 5 | 6 | 7 | 8+ |
| 1st | 1 | 1 | 1 | 1 | 1 | 1 | 1 | 1 |
| 2nd | 1 | 2 | 2 | 2 | 2 | 2 | 2 | 2 |
| 3rd | 1 | 1 | 3 | 3 | 3 | 3 | 3 | 3 |
| 4th | 1 | 2 | 1 | 4 | 4 | 4 | 4 | 4 |
| 5th | 1 | 1 | 2 | 1 | 5 | 5 | 5 | 5 |
| 6th | 1 | 2 | 3 | 2 | 1 | 6 | 6 | 6 |
| 7th | 1 | 1 | 1 | 3 | 2 | 1 | 7 | 7 |
| 8th | 1 | 2 | 2 | 4 | 3 | 2 | 1 | 8 |
| 9th | 1 | 1 | 3 | 1 | 4 | 3 | 2 | 1 |
| 10th | 1 | 2 | 1 | 2 | 5 | 4 | 3 | 2 |

1. Find out how many people are living in each household. Only count adults eligible for the survey, say, all those over 15.
2. Assign each eligible household member a number, starting with the youngest (#1). For example, let's say you were conducting a 10-house survey and you visited a four-person household consisting of mom, dad, a college-age son and a 90-year-old grandpa, grandpa would be #4.

3. Look up the column and row relevant to the household you're visiting; For example, if this is the 7th house you're surveying, and it is a four-person household, you'd look at column 7 row 4. If column 7 row 4 has the number 3, survey the person you've labeled number 3.

If the person chosen isn't available, find out when they might be met with and try again. Dad might be at work; in which case you would have to arrange to come again after the workday was over. If the interview still doesn't happen after 6 attempts, it should be abandoned as a casualty (Kapur, 2010).

## History of the Kish Grid

The Kish Grid was first proposed by Leslie Kish in 1949. It quickly became popular and was well-used for many years. It's become less effective because people today are less willing to give information — for example, a list of names and ages of everyone in their household— to a random telephone survey person. However, when that inhibition can be overcome, it is still a useful selection method.

## References

Kapur, 2010. Diaspora, Development, and Democracy: The Domestic Impact of International Migration from India. Princeton University.

Kish, L. (1949). A Procedure for objective respondent selection within a household. Journal of the American Sociological Association, 44, 380-387.

Lewis-Beck, M., Bryman, A., Liao, T. (2003) The SAGE Encyclopedia of Social Science Research Methods. SAGE Publications.

# Line Intercept Sampling

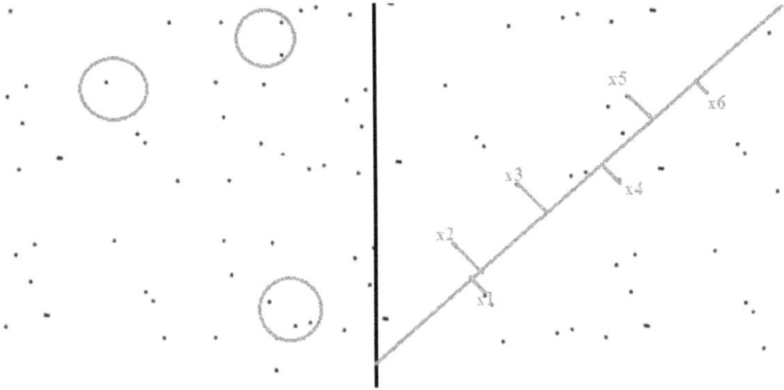


*Traditional sampling (left) and LIS (right).*

In **line-intercept sampling** (also known as line transect sampling*), an element is included in a sample from a particular region if a certain line segment (called a transect) intersects the element. One way to choose the transect is to select a baseline, then run the transect perpendicular to the baseline (Muttlak & Sadooghi-Alvandi, 1993). The technique is used in ecology to estimate percent vegetation coverage, coarse woody debris (CWD) coverage, and plant abundance.

Lines can be single, multiple, L-Shaped, or unequal length.

*Planet ecologists use line intercept sampling and line transect sampling to refer to the same sampling technique. While some authors (e.g., Pielou, 1985) insist on subtle differences between the two terms, the two terms are generally used synonymously (George & Valentine, 2002).

## General Steps

1. Take two measurements for each item of interest: intercept length and maximum width of the item perpendicular to the transect.
2. Measure the bare ground in the same way (Cox, 1990).
3. Repeat steps 1 & 2 if there are multiple items of interest (e.g., two plant species). Calculate the density for each item of interest: multiply the total reciprocals of maximum plant/vegetation/CWD by the unit area/total length of the transect (Cox, 1990).

## Advantages / Disadvantages

LIS takes less time than the quadrat method (which uses standard sized plots), but it's just as accurate (Fidelibus & Mac Aller, 1993).

One major disadvantage is that some of the LIS literature is "…misleading, or worse, incorrect"; misconceptions appear "distressingly often" (George & Valentine, 2002, p.263). Therefore, caution must be used when interpreting or replicating LIS statistical techniques from published papers. Kaiser (1983) is one noted exception.

It is usually hard to replicate the transect line more than twice, for financial, time or practical reasons. This results in difficulty estimating any parameter's variances (Muttlak & Sadooghi-Alvandi, 1993).

## References

Cox, G. Laboratory manual. of general ecology 6th Ed. Dubuque, Iowa: WIlliam C. Brown; 1990.

Fidelibus, M. & Mac Aller, R. (1993) Methods for Plant Sampling. Retrieved December 11, 2017 from: http://www.sci.sdsu.edu/SERG/techniques/mfps.html

George, T. & Valentine, H. (2003). LIS: Ell-shaped transects and multiple intersections. Environmental and Ecological Statistics 10. 263-279.

Kaiser, L. (1983) Unbiased estimation in line-interception sampling. Biometrics, 39, 965-76.

Muttlak, H. & Sadooghi-Alvandi, S. (1993). A Note on the Line Intercept Sampling Method. Biometrics. Vol. 49, No. 4 (Dec), pp. 1209-1215.

Pielou, E. (1975). Line intersect sampling. In Encyclopedia of Statistical Sciences. S. Kotz & N.L. Johnson (eds.). 5. Wiley, New York, pp. 70-74.

# Maximum Variation Sampling

**Maximum variation sampling** is what the name implies: a sample is chosen to ensure a wide variety of participants. Samples collected are typically small (from 3 up to about 50). Above 50 items, quota sampling or a similar non-probability method is simpler to implement and achieves better results.

## Why is it Used?

Reasons include:

- You want to understand how different groups of people view a specific topic. You know little about the population (and so find it difficult or impossible to get a random sample)
- Random sampling is otherwise not practical (because of logistics or a small population).
- You want your sample to be as representative as possible; by sampling the extremes, together they may represent an "average" respondent.

## Examples

- Researchers are conducting a door-to-door surveys to find attitudes towards single parents. During the day, they are more likely to encounter stay-at-home parents and retirees. Therefore, the researchers stagger survey hours for 8 a.m., 4 p.m. and 8 p.m. to include a wide variety of people.
- A researcher is investigating why people don't complete their prescribed course of antibiotics and thinks that socioeconomic class may be a reason. They survey one rich community, and one poor community.
- A researcher is studying the volunteer choices of college students. They include students from different socioeconomic backgrounds. The researcher also makes sure to include students of different race, nationality, culture, and work experiences.

Even when sampling from extremes (e.g., rich and poor, old and young), the population should otherwise be as homogeneous (similar) as possible. For example, if you collect data from rich urban people and poor backcountry dwellers, your sample is likely to be affected by other factors like access to services, public transportation, or job prospects.

**Reference:**

World Health Organization. Retrieved Jun 22, 2016 from http://apps.who.int/medicinedocs/en/d/Js6169e/7.3.html.

# Multistage Sampling

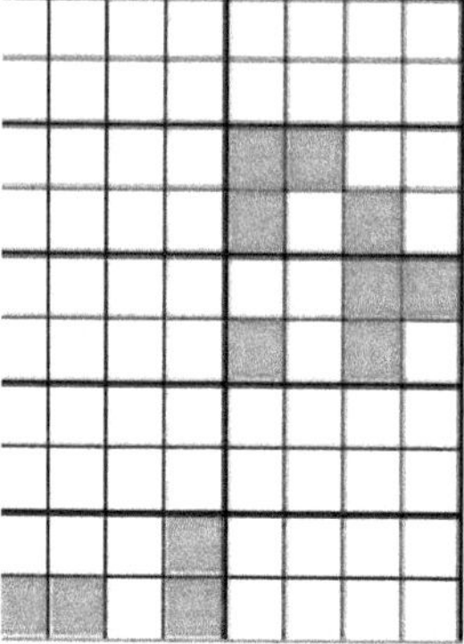

*Multistage sampling of 4 items from 3 blocks.*

**Multistage sampling** divides large populations into stages to make the sampling process more practical. A combination of stratified sampling or cluster sampling and simple random sampling is usually used.

Let's say you wanted to find out which subjects U.S. school children preferred. A population list—a list of all U.S. schoolchildren—would be near-impossible to come by, so you cannot take a sample of the population. Instead, you divide the population into states and take a simple random sample of states. For the next stage, you might take a simple random sample of schools from within those states. Finally, you could perform simple random sampling on the students within the schools to get your sample.

To classify multistage sampling as probability sampling, each stage must involve a probability sampling method.

## Advantages and Disadvantages

Multistage sampling is flexible, cost effective and easy to implement. You can use as many stages as you need to reduce the sample to a workable size, with no restrictions on how you divide the groups.

However, as the method has a subjective component, it has problems with **external validity** (how well your study outcome can be applied to other settings). It is also less accurate than simple random sampling.

## Real Life Examples of Multistage Sampling

- The Census Bureau uses multistage sampling for the U.S. National Center for Health Statistics' National Health Interview Survey (NHIS). A multistage probability sample of 42,000 households in 376 probability sampling units (PSUs are usually counties or groups of counties), which are chosen in groups of around four adjacent households.
- The U.S. based Gallup poll uses multistage sampling. For example, they might randomly choose a certain number of area codes then randomly sample several phone numbers from within each area code.
- Johnston et al.'s survey on drug use in high schools used three stage sampling: geographic areas, followed by high schools within those areas, followed by senior students in those schools.
- The Australian Bureau of Statistics divides cities into "collection districts", then blocks, then households. Each stage uses random sampling, creating a need to list specific households only after the final stage of sampling.

### References

Levine, D. (2014). Even You Can Learn Statistics and Analytics: An Easy to Understand Guide to Statistics and Analytics 3rd Edition. Pearson FT Press

McBurney, D. & White, T. (2009). Research Methods. Cengage Learning.

## Reservoir Sampling

**Reservoir sampling** is a quota-based random sampling method, used to get a particular sample size when you don't know the population size (e.g., when you're dealing with a data stream of unknown length). It can also be used to create a sample for very large data sets.

It's called reservoir sampling because the selected items are placed into a reservoir (a holding set). As each "streamtuple" is received, the algorithm updates dynamically. The reservoir can be updated with replacement, or without replacement.

Originally developed for one-pass processing from magnetic tapes (Andrade et al. 2014), reservoir sampling is now used for one pass stream processing in data mining.

### Reservoir Sampling Without Replacement

A reservoir sample without replacement is one where every distinct element has an equal probability of being selected:

$$p = 1/\binom{n}{m}$$

Where:

- n = population size.
- m = a distinct element.

As the sampling is done without replacement, each element in the set is distinct and is only selected once.

### Reservoir Sampling With Replacement

Reservoir sampling with replacement means that every element has the possibility of being chosen for the reservoir more than once. It should guarantee that every element in the sample has an equal chance (1/n) of being placed in a certain position in the sample, no matter what elements are in the other positions. Formally, this is written as:

$P(\mathcal{J} - \{i_1, i_2, \ldots, i_m\}) = 1/nm$

### References

Andrade, H. et al. (2014). Fundamentals of Stream Processing: Application Design, Systems, and Analytics. Cambridge University Press.

Park, B. et al. Reservoir-Based Random Sampling with Replacement from Data Stream. (1987). In Proceedings of the Fourth SIAM International Conference on Data Mining (Proceedings in Applied Mathematics) 4th ed. Edition. Society for Industrial and Applied Mathematics. pp. 492-496.

Vitter, J. (1985). Random Sampling with a Reservoir. ACM Transactions on Mathematical Software, Vol. 11, No. 1, March.

Steele, P. & Pallone, S. (2017). Reservoir Sampling. Retrieved January 6, 2021 from: https://people.orie.cornell.edu/snp32/orie_6125/algorithms/reservoir-sampling.html

## Respondent-Driven Sampling

**Respondent-driven sampling** can be thought of as a group of snowballs, each rolling in their own direction and growing larger.

Respondent-driven sampling (RDS) is similar to snowball sampling, a chain-referral sampling method where participants recommend other people they know. Both techniques are useful for sampling from hard-to-reach populations, like IV drug users or sex workers. The main difference between the two is that RDS is mathematically tweaked to add an element of randomness. RDS can be thought of as a group of snowballs, each rolling down a hill in their own random direction.

Snowball sampling is a non-probability sampling method that doesn't need a list of members to sample from (something that's hard or impossible to get for hard-to-reach populations). As there isn't a list of members, there's no way to get a random sample. RDS is a group of methods that convert chain-referral methods, like snowball sampling, into results that can be analyzed statistically. In other words, although it starts off as a non-probabilistic model, it ends up as a mixture between non-probabilistic and probabilistic methods. The probabilistic part of the method is a somewhat complex mathematical formula that compensates for the non-random collection method. Basically, the researchers:

1. Keep track of who referred who in the sample. The general idea behind RDS is if the "seeds" (the original members of the sample) generate enough random "sprouts" (i.e., referrals), then a large enough sampling will eliminate bias.
2. Create a model of the seeds and sprouts.
3. Weight the model to account for non-randomness.

## Problems with Respondent-Driven Sampling

RDS was first developed as part of an HIV-prevention study (Heckathorn). A series of later studies developed the methods seen today. As this is a new method, there isn't a clear consensus on what the results of RDS mean in the real world. Some authors (like McCreesh) found that bias wasn't eliminated from studies even when RDS methods had been correctly deployed. That said, RDS offers some benefits (like the potential elimination of bias) over other convenience methods.

**References:**

Heckathorn Douglas D. Respondent-Driven Sampling: A New Approach to The Study of Hidden Populations. Social Problems. 1997;44:174–199.

McCreesh, N et. al. Evaluation of respondent-driven sampling. Epidemiology. 2012 Jan;23(1):138-47. doi: 10.1097/EDE.0b013e31823ac17c.

---

## Random Sampling

A **random sample** is a sample that is chosen randomly. It could be more accurately called a *randomly chosen sample*. Random samples are used to avoid bias and other unwanted effects. Of course, it isn't quite as simple as it seems: choosing a random sample isn't as simple as just picking 100 people from 10,000 people. You must be sure that your random sample is truly random.

Note that the word "random" in random sample doesn't exactly meet the dictionary definition of the word. If you Google "define:random" then you'll read that it means:

> *...made, done, happening, or chosen without method or conscious decision. "a random sample of 100 households"*

It isn't true that a random sample is chosen "without method of conscious decision." A conscious decision must be made to use a method that will result in randomness. Simple random sampling is one way to choose a random sample.

### WHAT IS A SIMPLE RANDOM SAMPLE?

A **simple random sample** is often mentioned in elementary statistics classes, but it's one of the least used techniques. In theory, it's easy to understand. However, in practice it's tough to perform.

Technically, a simple random sample is a set of *n* objects in a population of N objects where all possible samples are equally likely to happen. Here's a basic example of how to get a simple random sample: put 100 numbered bingo balls into a bowl (this is the population N). Select 10 balls from the bowl without looking (this is your sample *n*). Note that it's important not to look as you could (unknowingly) bias the sample. While the "lottery bowl" method can work fine for smaller populations, you'll be dealing with much larger populations in real life situations.

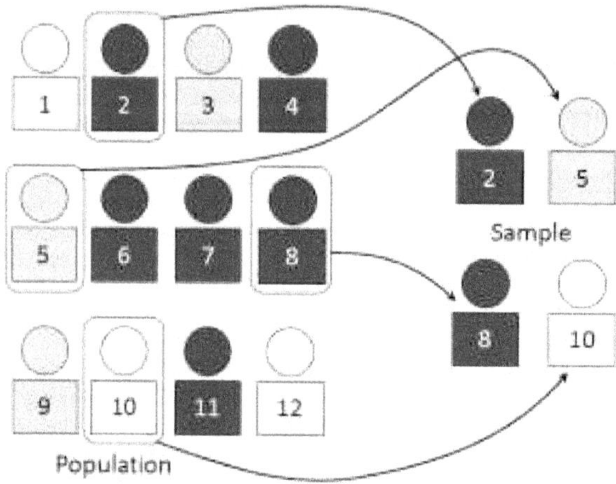


*Simple random sampling of a sample "n" of 3 from a population "N" of 12. Image: Dan Kernler / Wikimedia Commons*

Imagine the people illustrated in the image above are game pieces. Place the 12 game pieces in a bowl and (again, without looking) choose 3. This is simple random sampling.

A simple random sample is chosen in such a way that every set of individuals has an equal chance to be in the selected sample. It sounds easy, but SRS is often difficult to employ in surveys or experiments. In addition, it's very easy for bias to creep into samples obtained with simple random sampling. Sometimes it's impossible (either financially or timewise) to get a realistic sampling frame (the population from which the sample is to be chosen). For example, if you wanted to study all the adults in the U.S. who had high cholesterol, the list would be practically impossible to get unless you surveyed every person in the country. Therefore, other sampling methods would probably be better suited to that experiment.

The simplest example of SRS would be working with things like dice or cards — rolling the die or dealing cards from a deck can give you a simple random sample. But in real life you're usually dealing with people, not cards, and that can be a challenge.

## How to Perform Simple Random Sampling: Example

A larger population might be "All people who have had strokes in the United States." That list of participants would be extremely hard to obtain. Where would you get such a list in the first place? You could contact individual hospitals (of which there are thousands and thousands…) and ask for a list of patients (would they even supply you with that information? If you could somehow obtain this list, then you will end up with a list of 800,000 people which you then have to put into a "bowl" of some sort and choose random people for your sample. This type of situation is the type of real-life situation you'll come across and is what makes getting a simple random sample so hard to undertake.

*Example question*: Outline the steps for obtaining a simple random sample for outcomes of strokes in U.S. trauma hospitals.

Step 1: Make a list of all the trauma hospitals in the U.S. (there are several hundred: the CDC keeps a list).

Step 2: Assign a sequential number to each trauma center (1,2,3…n). This is your **sampling frame** (the list from which you draw your simple random sample).

Step 3: Figure out what your sample size is going to be.

Step 4: Use a **random number generator** (Excel has one built in) to select the sample, using your sampling frame (population size) from Step 2 and your sample size from Step 3. For example, if your sample size is 50 and your population is 500, generate 50 random numbers between 1 and 500.

Warning: If you compromise (say, by not including *all* trauma centers in your sampling frame), it could open your results to bias.

### Simple Random Sample vs. Random Sample

A **simple random sample** is like a random sample. The difference between the two is that with a simple random sample, each object in the population has an equal chance of being chosen. With random sampling, each object does not necessarily have an equal chance of being chosen. Unequal probability sampling isn't usually addressed in basic statistics courses.

## Stratified Random Sampling

**Stratified random sampling** is used when your population is divided into strata (characteristics like male and female or education level), and you want to include the stratum when taking your sample. The stratum may be already defined (like census data) or you might make the stratum yourself to t the purposes of your research.

Stratified random sampling is very similar to random sampling. However, these samples are more difficult to create as you must have detailed information about what categories your population falls into.

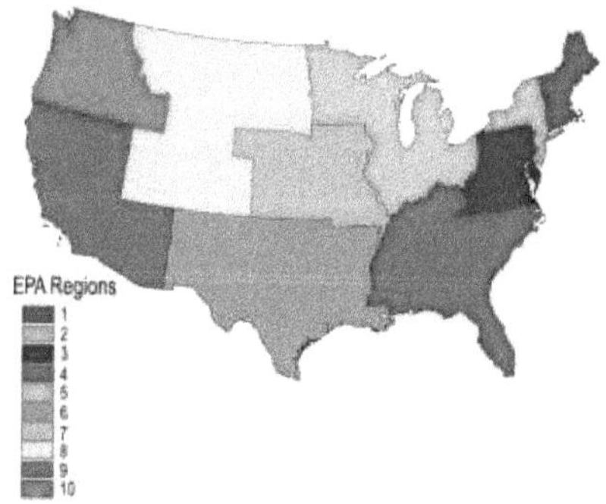


*The stratum in this map are defined by EPA region. Image: USGS.*

### How to Perform Stratified Random Sampling

To perform stratified random sampling, take a random sample from within each category or stratum. Let's say you have a population divided into the following strata:

- Category 1: Low socioeconomic status — 39 percent.
- Category 2: Middle class — 38 percent.
- Category 3: Upper income — 23 percent.

To get the stratified random sample, you would randomly sample the categories so that your eventual sample size has 39 percent of participants taken from category 1, 38 percent from category 2 and 23 percent from category 3. What you end up with is a mini representation of your population. According to University of California at Davis, the following steps should be taken to obtain the stratified sample:

1. Name the target population.
2. Name the categories (stratum) in the population.
3. Figure out what sample size you need.
4. List all the cases within each stratum.
5. Make a **decision rule** to select cases. A decision rule tells you how the items will be chosen. For example, you might select the items using the largest set of random numbers.
6. Assign a random number to each case.
7. Sort each case by random number.
8. Follow your decision rule (#5 above) to choose your participants.

Stratified random sampling for larger data sets is usually performed using statistical software.

## How to Get a Stratified Random Sample: Example

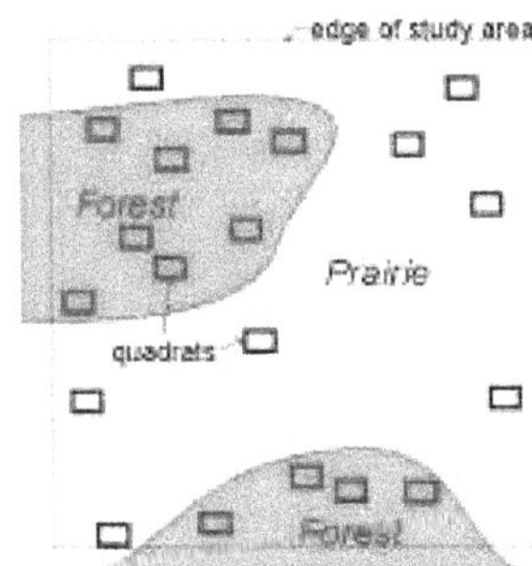

*Stratified random sampling is useful when you can subdivide areas. Image: Oregon State*

"Stratified" means "in layers," so to get a stratified random sample you first need to make the layers. What layers you have depends on characteristics of your population. For example, if you are surveying U.S. residents about their plans for retirement, you might want your layers to represent different age groups. The sample size for each strata (layer) is proportional to the size of the layer:

*Sample size of the strata = size of entire sample / population size * layer size.*

## How to Get a Stratified Random Sample: Steps

Example question: You work for a small company of 1,000 people and want to find out how they are saving for retirement. Use stratified random sampling to obtain your sample.

Step 1: Decide how you want to stratify (divide up) your population. For example, people in their twenties might have different saving strategies than people in their fifties.

Step 2: Make a table representing your strata. The following table shows age groups and how many people in the population are in that strata:

| Age | Total Number of People in Strata |
|---|---|
| 20-29 | 160 |
| 30-39 | 220 |
| 40-49 | 240 |
| 50-59 | 200 |
| 60+ | 180 |

Step 3: Decide on your sample size. For this example, we'll assume your sample size is 50.

Step 4: Use the stratified sample formula (Sample size of the strata = size of entire sample / population size * layer size) to calculate the proportion of people from each group:

| Age | Total Number of People in Strata | People in Sample |
|---|---|---|
| 20-29 | 160 | 50/1000 * 160 = 8 |
| 30-39 | 220 | 50/1000 * 220 = 11 |
| 40-49 | 240 | 50/1000 * 240 = 12 |
| 50-59 | 200 | 50/1000 * 200 = 10 |
| 60+ | 180 | 50/1000 * 180 = 9 |

Note that all the individual results from the stratum add up to your sample size of 50: 8 + 11 + 12 + 10 + 9 = 50

Step 5: Perform random sampling (e.g., simple random sampling) in each stratum to select your survey participants.

Tip: Each element in your population should only fit into one stratum. In other words, one person cannot be in more than one group.

**Stratified Randomization**

*AABBAABB*
*BBBBAAA*
*AABAAABB*

*Stratified randomization uses permuted blocks within strata.*

In **stratified randomization** (sometimes called Stratified Permuted Block Randomization), trial participants are subdivided into strata, then permuted block randomization is used for each stratum. **Permuted block randomization** is a way to randomly allocate a participant to a treatment group, while maintaining a balance across treatment groups. Each "block" has a specified number of randomly ordered treatment assignment The goal is to create a balance of clinical/prognostic factors because the trial may not have valid results if factors are not well balanced.

This form of randomization is only important for small clinical trials (fewer than 400 patients), where known clinical factors (like gender, age, disease stage, or obesity) are thought to effect treatment outcomes. Larger trials don't use the technique because it's unlikely to find imbalances in clinical factors for a randomized large group.

Don't try to balance every clinical factor — choose the most important ones. Too many strata can result in too few patients in each. As an extreme, you could end up with only one patient — or even zero patients — in each strata. Therefore, you should keep strata to a minimum. Between 1 and 5 factors (randomization variables) is commonly recommended; Each factor usually has between 2 and 4 levels. The recommended number of stratification factors is usually one or two.

The number of patients in each strata do not have to be even. The important thing is to make sure an equal number of patients are assigned to block A or B, which is taken care of if you use permuted block randomization within the strata.

## Difference Between Cluster and Stratified Sampling

For a **stratified random sample**, a population is divided into stratum, or sub-populations, before sampling. At first glance, the two techniques seem very similar. However, in cluster sampling the actual cluster is the sampling unit; in stratified sampling, analysis is done on elements within each strata. In cluster sampling, a researcher will only study selected clusters; with stratified sampling, a random sample is drawn from each strata.

## References

Polit DF Beck CT (2012). Nursing Research: Generating and Assessing Evidence for Nursing Practice, 9th ed. Philadelphia, USA: Wolters Klower Health, Lippincott Williams & Wilkins.

Therneau, T. How many Stratification Factors is "Too Many" to Use in a Randomization Plan? Retrieved July 21, 2016 from http://www.mayo.edu/research/documents/biostat-57pdf/doc-10027486.

# Sequential Sampling

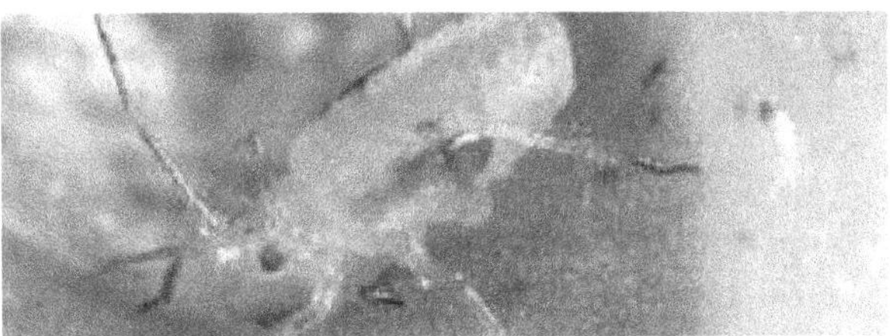

*Sequential sampling is often used in fields like Integrated Pest Management.*

In **sequential sampling**, a sequence of one or more samples is taken from a group. Once the group has been sampled, a statistical test is performed (usually, with software) to see if you can reach a conclusion. If you can't, the whole procedure is repeated. A characteristic feature of sequential sampling is that the sample size is not set in advance, because you don't know at the outset how many times you'll be repeating the process.

This method is designed for two clear choices. For example:

- Is the heat in a system above or below a critical level? Heat is measured in one part of the system to see if it has reached the critical level. If the heat is close to the critical level, but not over it, resample and repeat the calculations.
- Should I spray pesticide or not? Pests could be counted on a plant. If there are many pests, spray pesticide. If there are a small number of pests, do not spray pesticide. If there are a middling number of pests, sample another plant.

Sequential samples can either be:

- Item-by-item: one sample at a time.
- Group: sample sizes of two or more.

To use this method, you must be able to sample serially. If you have to choose all of your sample items at the same time, you should choose another sampling method (like simple random sampling or a non-probability sampling method).

## Three Outcomes

With traditional sampling methods, a statistical test has one of two possible results: you either **reject the null hypothesis** (replacing it with your hypothesis), or you do not.

The **null hypothesis**, $H_0$ is the commonly accepted fact; it is the opposite of the **alternate hypothesis**. Researchers work to reject, nullify, or disprove the null hypothesis. Researchers come up with an alternate hypothesis, one that they think explains a phenomenon, and then work to reject the null hypothesis.

With sequential sampling, you have three possibilities:

- Reject the null hypothesis (end the experiment).
- Do not reject the null hypothesis (end the experiment).
- Fail to draw any conclusion (draw another sample and repeat the test).

## Time-Sequential Sampling

In this variant, sometimes called **time-sequential classification**, you use time as your sampling frame instead of a physical population to sample from. For example, you might choose a sample member at 24-hour intervals.

## Advantages and Disadvantages

Although it sounds like the process could go on and on forever, sequential sampling usually ends up with smaller samples than traditional (set size) sampling. However, the mathematics needed to analyze data for sequential sampling is much more complex and the procedure is generally more time consuming (and can be more expensive) than fixed-size sampling. Computers are often required to develop sequential sampling plans. To simplify the process, many elds have developed sampling plans based on the Poisson distribution, negative binomial distributions, and others to maximize efficiency. For example, Pedigo and Buntin (1993, p.578) list dozens of plans for different agricultural arthropods.

**References:**

Krebs, C. (2014). Ecological Methodology. 3rd. Ed.
Pedigo, L. & Buntin, G. (1992). Handbook of Sampling Methods for Arthropods in Agriculture. CRC Press.

## Snowball Sampling

**Snowball sampling** is where research participants recruit other participants for a test or study. It is used where potential participants are hard to find. It's called snowball sampling because (in theory) once you have the ball rolling, it picks up more "snow" along the way and becomes larger and larger. Snowball sampling is a non-probability sampling method. It doesn't have the probability involved, with say, simple random sampling (where the odds are the same for any participant being chosen). Rather, the researchers used their own judgment to choose participants.

Snowball sampling consists of two steps:

1. Identify potential subjects in the population. Often, only one or two subjects can be found initially.
2. Ask those subjects to recruit other people (and then ask those people to recruit. Participants should be made aware that they do not have to provide any other names.

These steps are repeated until the needed sample size is found. Ethically, the study participants should not be asked to identify other potential participants. Rather, they should be asked to encourage others to come forward.

When individuals are named, it's sometimes called "cold-calling", as you are calling out of the blue. Cold calling is usually reserved for snowball sampling where there's no risk of potential embarrassment or other ethical dilemmas. For example, it would be easier to cold-call participants in a study for families who regularly dine at fast-food restaurants than it would be to cold-call people who are having extra-marital affairs.

Snowball sampling can be a tricky ethical path to navigate. Therefore, you'll probably be in contact with an institutional review board, or another department similarly involved in ethics.

### Why is Snowball Sampling Used?

Some people may not want to be found. For example, if a study was investigating cheating on exams, shoplifting, drug use, prostitution, or any other "unacceptable" societal behavior, potential participants would be wary of coming forward because of possible rami cations. However, other study participants would likely know other people in the same situation as themselves and could inform others about the benefits of the study and reassure them of confidentiality.

### Advantages and Disadvantages of Snowball Sampling

Advantages:

- It allows for studies to take place where otherwise it might be impossible to conduct because of a lack of participants.
- Snowball sampling may help you discover characteristics about a population that you weren't aware existed. For example, the casual illegal downloader vs. the for-profit downloader.

Disadvantages:

- It is usually impossible to determine the sampling error or make inferences about populations based on the obtained sample.
- Snowball sampling is also known as cold-calling, chain sampling, chain-referral sampling, and referral sampling.

## Square Root Biased Sampling

**Square root biased sampling** isn't a technique that's widely used. That said it is an interesting technique that attempts to address the problem of profiling at airport screenings.

Statisticians strive to choose random people for surveys and experiments. This random sampling doesn't happen at airport screenings, presumably because people who "look" a certain way are more likely to be terrorists. This is a problem William H. Press attempts to address with square root biased sampling. He states:

*"...resources are wasted on the repeated screening of higher probability, but innocent, individuals."*

In other words, profiling by ethnicity, having the same name as someone on a watch list isn't a mathematically sound way to catch a terrorist.

Square root biased sampling adds simple random sampling to profiling. Simple random sampling is where individuals are chosen completely by chance from a population. The addition of SRS increases the chance a guilty person will be found. It should also mean innocent travelers are more likely to breeze through security. The system works by assigning the same profiling. Instead of a profiled passenger being selected for screening every time, they may be pulled aside less frequently. For example, if a person is 10 times more likely to be a terrorist, the current system would pull them aside ten times more often than a non-pro led passenger. This basically means every time that profiled person travels, they will be pulled aside. The addition of SRS means that the passenger will only be pulled aside three times as often.

## Systematic Sampling

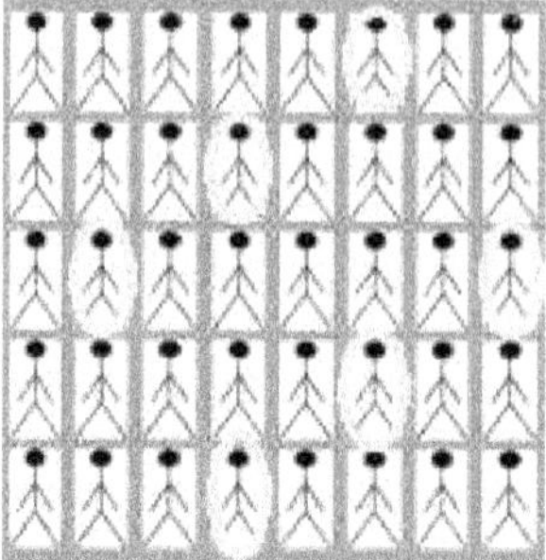

*A systematic sample where every 6th person is chosen (highlighted in yellow).*

One way to get a fair and random sample is to assign a number to every population member and then choose the nth member from that population. For example, you could choose every 10th member, or every 100th member. This method of choosing the nth member is called **systematic sampling**.

When you're sampling from a population, you want to make sure you're getting a fair representation of that population. Otherwise, your statistics will be biased or **skewed** (i.e., off center or not symmetric) and perhaps meaningless.

Systematic sampling is quick and convenient when you have a complete list of the members of your population (for example, the members of Congress). However, if there's a pattern to the original list, then bias may creep into your statistics.

For example, if a list of people is ordered as MFMFMFMF, then choosing every 10th number will give you a sample consisting entirely of females. You could randomly shuffle the list before choosing the nth item or you could use repeated systematic sampling, where you take several small samples from the same population. It's used if you aren't sure you have a completely random list and you want to avoid bias.

## How to Perform Systematic Sampling: Steps

Step 1: Assign a number to every element in your population. For this simple example, let's say you have a population of 100 people, so you'll assign the numbers 1 to 100 to the group.

Step 2: Decide how large your sample size should be. For this example, let's say you need a sample of 10 people.

Step 3: Divide the population by your sample size. For this example, your population is 100 and your sample size is 10, so:

100 10 = 10

This is your "nth" sampling digit (for this example, you'll choose every 10th item):

1 2 3 4 5 6 7 8 9 **10** 11 12 13 14 15 16 17 18 19 **20** 21 22 23 24 25 26 27 28 29 **30** 31 32 33 34 35 36 37 38 39 **40** 41 42 43 44 45 46 47 48 49 **50** 51 52 53 54 55 56 57 58 59 **60** 61 62 63 64 65 66 67 68 69 **70** 71 72 73 74 75 76 77 78 79 **80** 81 82 83 84 85 86 87 88 89 **90** 91 92 93 94 95 96 97 98 99 **100.**

## Repeated Systematic Sampling

The first three steps are the same:

Step 1: Assign a number to every element in your population.

Step 2: Decide how large your sample size should be.

Step 3: Divide the population by your sample size. For example, if your population is 100 and your sample size is 10, then:

100 / 10 = 10

This is your "nth" sampling digit.

The main difference is that you'll stop part way through the sampling and switch. Follow the steps to see how this is done:

### Sampling from Group 1

Step 4: Use the sampling digit from Step 3 up to a certain point. This is usually a judgment call; Exactly where you stop is usually quite arbitrary. The goal is to divide your population into parts. For this example, we'll sample up to 50; By stopping at 50 we are splitting the entire group into two sections.

First, you'll sample from the first half of the group (in Step 5, you'll sample from the remainder).

1 2 3 4 5 6 7 8 9 **10** 11 12 13 14 15 16 17 18 19 **20** 21 22 23 24 25 26 27 28 29 **30** 31 32 33 34 35 36 37 38 39 **40** 41 42 43 44 45 46 47 48 49 **50**

### Systematic Sampling from Group 2

Step 5: Switch to a different starting point and then continue sampling with the nth digit. Again, this is usually a judgment call. For this example, we'll switch from 50 to 51.

**61** 62 63 64 65 66 67 68 69 70 **71** 72 73 74 75 76 77 78 79 80 **81** 82 83 84 85 86 87 88 89 90 **91** 92 93 94 95 96 97 98 99 100.

Note that we only have 9 in our sample (we wanted 10), so return to the beginning of the list and continue: **1** 2 3 4 5 6 7 8 9 10. We now have ten items.

### References

Ken Black (2004). Business Statistics for Contemporary Decision Making (Fourth (Wiley Student Edition for India) ed.). Wiley-India.

## Total Population Sampling

**Total population sampling** is a type of purposive sampling where the whole population of interest (i.e., a group whose members all share a given characteristic) is studied. A **purposive sample** is where a researcher selects a sample based on their knowledge about the study and population. The participants are chosen based on the purpose of the sample, hence the name.

It is most practical when the total population is of manageable size, such as a well-defined subgroup of a larger population. For example, total population sampling would be a good way to conduct a survey meant to get the opinions of third-grade schoolgirls from a local elementary school; all that would be required is to survey each girl in every third-grade classroom.

In practice, total population sampling is done when the target group is small and set apart by an unusual and well-defined characteristic. This might be something like US city mayors that have immigrant backgrounds, or college presidents with more than five children, or family members of those suffering from a rare disease that affects one in a million.

## Steps in Total Population Sampling

1. Completely define the population and the characteristic or characteristics which set it apart.
2. Create a list of the population.
3. Collect relevant data from all members on the list.

## Advantages and Disadvantages

Gleaning information from the total population often gives deeper insights into a than partial target population samples would be capable of. It has the potential to allow a researcher to paint a much more complete picture, and greatly reduces guesswork.

It also eliminates the risk of biased sample selection that is often encountered in would-be random study samples.

But since sampling a total population means you need to first draw up a list of the whole population, it is often not easy. That first list is, in many cases, very difficult or almost impossible to get. This may be the most time-consuming part of a study; and any errors or omissions can aw the whole study.

Another disadvantage is the challenge, even after the complete list is made, in ensuring the data collected really is complete. In the case of a survey, non-response bias (where certain members of the population are less likely or more likely to respond) can badly skew results if the non-respondents are not random members of the population; if they share certain characteristics not shared by others.

### References


Lavrakas, P. (2008). Encyclopedia of Survey Research Methods 1st Edition. SAGE Publications.

Laerd Dissertation, Lund Research Ltd. Total Population Sampling. Retrieved from http://dissertation.laerd.com/total-population-sampling.php on April 18, 2018

Crossman, Ashley. Understanding Purposive Sampling: An Overview of the Method and Its Applications. Retrieved from https://www.thoughtco.com/purposive-sampling-3026727 on April 18, 2018


# Typical Case Sampling

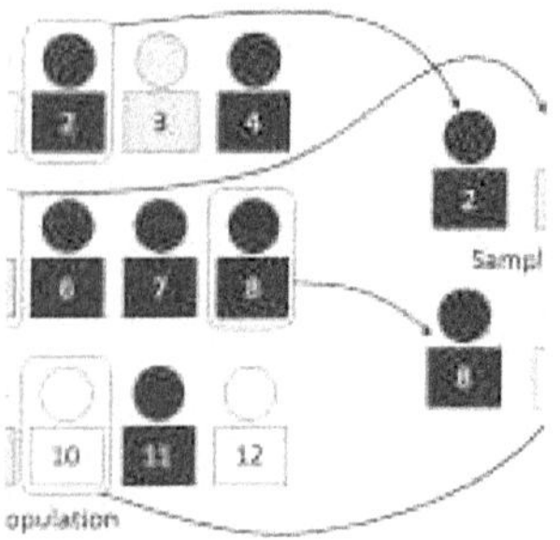

**Typical Case Sampling** allows you to develop a profile about what is usual or average for a particular phenomenon. For example, let's say you were studying violence in schools. The first step would be to list all the criteria that de ne violence for a "typical" school. Then you would choose schools that meet that criteria. You would want to select schools that are "average" (meeting your selected criteria) instead of schools with very high or very low violence rates.

"Typical" in non-probability sampling doesn't have the same meaning as a typical result from probability sampling methods like simple random sampling. In probability sampling, a typical result would be one that reflected the average result found in the whole population. For example, if your population weighed around 200 pounds, then your typical member would also weight about 200 pounds. In non-probability sampling, a typical case means that you can use your findings to cross-compare between samples — or cross check your results with similar studies (as opposed to comparing your results with what would be expected in a population).

## Limitations

Like most non-probability sampling methods, typical case sampling is less than ideal and is usually used when there are severe limitations on time or resources (Henry, 1990). This type of sampling is typically subjected to scrutiny, so you should take steps to avoid pitfalls like selection bias or minimize their impact. For example, a simple way to minimize selection bias is to avoid including any extreme cases in your sample.

Another limitation with this type of sampling is that you may not be able to deduce what a "typical" case may be.

One solution is to ask as many experts in the field as possible what they would consider to be typical cases. Another option is to use another sampling technique — like maximum variation sampling — to identify typical cases prior to choosing cases for your study (Baran, 2016).

You could also use demographic data to identify your cases. Wolcott (1994) used a National Education Survey, which reported the typical school principal was "a male, married, between the ages of 35 and 49, has had 10-19 years total experience in schools, and was an elementary classroom teacher" (Wolcott, 1994, p.117); He then looked for a person fitting that description.

## References

Baran, M. (2016). Mixed Methods Research for Improved Scientific Study. IGI Global.
Henry, G. (1990). Practical Sampling. SAGE publications.
Wolcott, H. (1994). Transforming Qualitative Data. Description, Analysis, and Interpretation. SAGE publications.

# 3. CONDITIONS AND PITFALLS

## 10% Condition

The **10% condition** states that sample sizes should be no more than 10% of the population. Whenever samples are involved in statistics, check the condition to ensure you have sound results. Some statisticians argue that a 5% condition is better than 10% if you want to use a standard normal model.

For example, the 10% condition usually applies when you:

- Draw samples without replacement in the **Central Limit Theorem.** The CLT states that the sampling distribution of the sample means approaches a normal distribution as the sample size gets larger —no matter what the shape of the population distribution. All this is saying is that as you take more samples, especially large ones, your graph of the sample means will look more like a bell shape.
- Have proportions from two groups.
- Check differences of means for very small populations or an extremely large sample.
- Use student's-t test (a statistical test to find differences in means).
- Are dealing with **Bernoulli trials** (i.e., trials with YES/NO or SUCCESS/FAILURE) that are not independent events. Normally, Bernoulli trials *are* independent, but it's okay to violate that rule if the sample size is less than 10% of the population.

The 10% condition isn't normally checked for:

- Chi-square tests (another hypothesis test, which you can use to test differences in groups).
- Differences of means (except for small populations or for extremely large samples).

- Randomized experiments (there is no sampling in randomized experiments, so the 10% condition can't be used).

Usually, you won't find the 10% condition mentioned for statistical means. When you make inferences about proportions, the 10% condition is necessary because of the large samples. But for means, the samples are usually smaller, making the condition necessary only if you are sampling from a very small population.

The condition applies in Bernoulli trials because in most cases you sample without replacement. For example, in a telephone survey asking "yes" or "no," you don't put a person who has already answered the question back in the pool.

## Large Enough Sample Condition

The **Large Enough Sample Condition** tests whether you have a large enough sample size compared to the population. A general rule of thumb for the Large Enough Sample Condition is that $n \geq 30$, where n is your sample size. However, it depends on what you are trying to accomplish and what you know about the distribution. In general, the Large Enough Sample Condition applies if any of these conditions are true: You have a symmetric distribution or unimodal (single-peaked) distribution without outliers: a sample size of 15 is "large enough."

- You have a moderately skewed distribution, that's unimodal without outliers; If your sample size is between 16 and 40, it's "large enough."
- Your sample size is >40 if you do not have outliers.
- Your population has a normal distribution.

### Central Limit Theorem and the Large Enough Sample Rule

The central limit theorem states that if a sample size is large enough, the distribution will be approximately normal. The general rule of $n \geq 30$ applies.

### Chi Square and the Large Enough Sample Condition

There are three different tests that use the chi-square; in each test, the assumptions and conditions are the same, including the Large Enough Sample Condition.

To know if your sample is large enough to use chi-square, you must check the Expected Counts Condition: if the count in every cell (the cells may be in a spreadsheet like Excel) is 5 or more, the cells meet the Expected Counts Condition, and your sample is large enough. Note that 5 is arbitrary and is open to interpretation. Some texts suggest that it's okay to have a few expected counts less than 5 (no more than 20%) if none are less than 1 (Yates, Moore & McCabe, The Practice of Statistics, 1999).

### Calculating sample size

Calculating sample size can be one of the most confusing aspects of statistics, mostly because of all the rules (and rules of thumb) surrounding appropriate sizes for different distributions. You must make sure your sample is sufficiently large, but not *too* large.

## Effective Sample Size

An **effective sample size** (sometimes called an adequate sample size) in a study is one that will find a statistically significant effect for a scientifically significant event. In other words, an effective sample size ensures that an important research question gets answered correctly. To achieve this, your sample must be the "right" size: neither too big nor too small. This is more of an art than a science.

Note: The term "effective sample size" is also used in the calculation of design effects and has a much narrower definition; specifically, it's the sample size you would expect if you used simple random sampling.

### Effect Sizes

An effective sample size is partially dependent on what **effect size** you're willing to work with. Let's say you're comparing two drugs A and B. You know that drug B is better than A. Drug B could be ten times better, or it could be slightly better. This variability (twice as much? ten times as much?) is what is called an effect size. In general, gains in your ability to detect effect size and increase statistical power (the "strength" of a test) is increases with larger sample sizes.

Having smaller effect sizes (i.e., the ability to detect smaller significant differences) is preferable. For example, drugs A and B might be for cholesterol, and drug B decreases cholesterol levels by 10 mg/dL. A sample size of 20 might detect changes in cholesterol of 20 mg/dL with a 90% power, while a sample size of 80 might detect changes of 10 mg/dL with the same power. The obvious choice would be to go with the test with the better effect size (the one that would detect smaller changes of 10 mg/dL). However, this comes with an important trade off: halving the value of an effect size will generally quadruple the sample size.

### Cost and Time Factors

Larger sample sizes involve more cost and time. Studies usually have limited budgets, which means you may not be able to get an ideal effective sample size. If you are limited in your sample size, changes in other aspects of your study design, like changing your data collection method, may result in a smaller effect size with the same sample size.

### A Note on Significance

You'll also want to consider whether a small change (like a 10 mg/dL change in cholesterol) has any real practical benefit. Normal cholesterol levels are 200 mg/dL while "high" is anything above 240 mg/dL. A drug that lowers cholesterol by 50 mg/dL would therefore be significant, while one that lowers it by 10 mg/dL is not. Therefore, although a sample size of 80 results in the desired effect size, it isn't going to be practically significant and is therefore not an effective sample size.

## Finite Population Correction Factor

The **Finite Population Correction Factor** (FPC) is used when you sample without replacement from more than 5% of a finite population. It's needed because under these circumstances, the Central Limit Theorem doesn't hold and the **standard error** of the estimate (which tells you how far your sample statistic deviates from the population parameter) will be too big. In basic terms, the FPC captures the difference between sampling with replacement and sampling without replacement.

Most real-life surveys involve finite populations sampled without replacement. For example, you might perform a telephone survey of 10,000 people; once a person has been called, they won't be called again.

Note: A downside of using the FPC is that it can cause uncertainty when applying the results to a larger population, so you should be careful when making inferences.

## Formula

The general formula is: FPC = ((N-n)/(N-1)) $^{1/2}$

Where:

- N = Population size.
- n = Sample size.

If the calculated value for the FPC is close to 1, it can be ignored. As the sample size falls under 5%, the value becomes somewhat insignificant (an FPC is .998 for a sample of 50).

This table of values shows how the FPC decreases for a population of 10,000 as the sample size gets larger.

| Sample Size | FPC |
|---|---|
| 1 | 1.000 |
| 10 | 1.000 |
| 25 | 0.999 |
| 50 | 0.998 |
| 100 | 0.995 |
| 500 | 0.975 |
| 1000 | 0.949 |
| 5000 | 0.707 |
| 8000 | 0.447 |

## How to Use the Formula

Place the correction at the end of the formula you want to use. For example, the standard error of the mean formula is:

$$\sigma_{\bar{x}} = \frac{\sigma}{\sqrt{n}}$$

And with the correction, the formula is.

$$\sigma_{\bar{x}} = \frac{\sigma}{\sqrt{n}} \sqrt{\frac{N - n}{N - 1}}$$

Or, for a confidence interval for a mean and unknown population standard deviation, the formula (with FPC) is:

$$\bar{X} \pm t_{\alpha/2} \frac{S}{\sqrt{n}} \sqrt{\frac{N - n}{N - 1}}$$

**Example**

Thirty people from a population of 300 were asked how much they had in savings. The sample mean($\bar{x}$) was $1,500, with a sample standard deviation of $89.55. Construct a 95% confidence interval estimate for the population mean.

**Solution** (using **degrees of freedom** = n – 1 = 29) and $t_{\alpha/2}$ = 2.0452 for a 95% **confidence level**):

$$\bar{x} \pm t_{\alpha/2} \frac{S}{\sqrt{n}} \sqrt{\frac{N - n}{N - 1}}$$

$$= \quad \$1,500 \pm (2.0452) \frac{89.55}{\sqrt{30}} \sqrt{\frac{300 - 30}{300 - 1}}$$

= $1,500 ± 33.44(0.9503)
= $1,500 ± 31.776
= $1,468.22 ≤ μ ≤ $1,531.78.

Note: The values for t can be found in the t-table: https://www.statisticshowto.com/tables/t-distribution-table/

**References:**

Kandethody, M. et, al. Mathematical Statistics with Applications. Elsevier India (2012). p.187.

# Oversampling

In statistics, **oversampling** involves taking higher, disproportionate samples than would otherwise be collected with random sampling. Depending on the structure of the survey or poll, oversampling might result in bringing low numbers of an underrepresented minority group to a representative proportion. In some cases, it might result in the minorities being over-represented. However, these "disproportionate" samples are aimed at effectively studying small groups, not biasing survey or poll results (Merger 2016).

## Benefits of Oversampling

Benefits include:

- Level the playing field and make a survey or sampling design more representative of a population.
- Reduce bias.
- Arguably, improve **validity**--the extent to which your research is well-founded and likely corresponds accurately to the real world); Not everyone agrees that oversampling results in improved validity (Couper, as cited in Lazar et. al).

Having a large response from a certain segment of the population reduces the likelihood of that population being underrepresented. For example:

- A 1992 telephone survey (Mohadjer & West) doubled the samples in geographic areas with large proportions of black and Hispanic residents to improve reliability estimates for these populations.
- In a survey on sexual assault, Busch et. al (2003) oversampled African Americans and Hispanics to match Texas's overall demographics. Sexual violence against black women routinely goes unreported (Now.org, 2019).

## References

Busch, N. et. al (2003). The Health Survey of Texans; A Focus on Sexual Assault.
Lazar, J. et al. (2017). Research Methods in Human-Computer Interaction 2nd Edition. Morgan Kaufmann.
Merger, A. (2016). Oversampling is used to study small groups, not bias poll results.
Mohadjer, L. & West, J. (1992). Effectiveness of Oversampling Blacks and Hispanics in the NHES Field Test: National Household Education Survey. U.S. Department of Education, Of ce of Educational Research and Improvement, National Center for Education Statistics.
Now.Org. Black Women and Sexual Violence.

# Sampling Error

Errors happen when you take a sample from the population rather than using the entire population. In other words, **sampling error** is the difference between the statistic you measure and the parameter you would find if you took a census of the entire population.

If you were to survey the entire population (like the US Census), there would be no error. It's nearly impossible to calculate the error margin. However, when you take samples at random, you estimate the error and call it the **margin of error**.

For example, if you wanted to figure out how many people out of a thousand were under 18, and you came up with 19.357%. If the actual percentage equals 19.300%, the difference (19.357 – 19.300) of 0.57 or 3% = the margin of error. If you continued to take samples of 1,000 people, you'd probably get slightly different statistics, 19.1%, 18.9%, 19.5% etc., but they would all be around the same figure. This is one of the reasons that you'll often see sample sizes of 1,000 or 1,500 in surveys: they produce a very acceptable margin of error of about 3%.

Formula: the formula for the margin of error is $1/\sqrt{n}$, where n is the size of the sample. For example, a random sample of 1,000 has about a $1/\sqrt{n}$; = 3.2% error.

Sample error can only be reduced, this is because it is an acceptable tradeoff to avoid measuring the entire population. In general, the larger the sample, the smaller the margin of error. There is a notable exception: if you use cluster sampling, this may increase the error because of the similarities between cluster members. A carefully designed experiment or survey can also reduce error.

## Another Type of Error

The non-sampling error could be one reason as to why there's a difference between the sample and the population. This is due to poor data collection methods (like faulty instruments or inaccurate data recording), selection bias, non-response bias (where individuals don't want to or can't respond to a survey), or other mistakes in collecting the data. Increasing the sample size will not reduce these errors. They key is to avoid making the errors in the first place with a well-planned design for the survey or experiment.

# 4. ADVANCED SAMPLING

## Undersampling

**Undersampling** attempts to reduce the bias (error) associated with imbalanced classes of data. In machine learning, undersampling and oversampling are two techniques that deal with imbalances in a training set (the part of data used to t a model). You can undersample the majority class, oversample the minority class, or combine the two techniques.

In general, undersampling (instead of oversampling) the majority class works best for large data sets. That's because with oversampling, you're adding more data points, which can lead to a data set that's too massive to use classifiers like support vector machines (García-Pedrajas, 2010).

### Random Undersampling

With random undersampling, you randomly remove members of the majority class until you reach a preset threshold.

One advantage to random selection here is that you don't have to make decisions on which points are important and which are not: you simply let the random process do the work. Several studies have shown that random selection performs as well as, if not better than, processes where deliberate removal choices are made.

However, a distinct disadvantage is that the process could remove important members. Problems tend to result in data that is non-smooth, has boundaries or small features (Dey, n.d.). One way to avoid this pitfall is to combine undersampling and boosting (Liu et al., as cited in García-Pedrajas, 2010). You might also want to manually resample or repair any holes in the data algorithmically.

### References

Dey, T. Undersampling and Oversampling in Sample Based Shape Modeling. Retrieved December 16, 2019 from: https://graphics.stanford.edu/courses/cs468-03-fall/Papers/deygiesen_undersampling.pdf

García-Pedrajas, N. et al. (2010). Trends in Applied Intelligent Systems: 23rd International Conference on Industrial Engineering and Other Applications of Applied Intelligent Systems, IEA/AIE 2010, Cordoba, Spain, June 1-4, 2010, Proceedings. Springer Science & Business Media.

# Resampling

**Resampling** techniques are a set of methods to either repeat sampling from a given sample or population, or a way to estimate the precision of a statistic.

Informally, resample can mean something a little simpler: repeat any sampling method. For example, if you're conducting a Sequential Probability Ratio Test and don't come to a conclusion, then you resample and rerun the test. For most intents and purposes though, if you read about resampling (as opposed to "resample"), then the author is most likely talking about a specific resampling technique.

The main techniques are:

1. Bootstrapping and Normal resampling (sampling from a normal distribution).
2. Permutation Resampling (also called Rearrangements or Rerandomization),
3. Cross Validation.

## 1. Bootstrapping and Normal Resampling

**Bootstrapping** is a type of resampling where large numbers of smaller samples of the same size are repeatedly drawn, with replacement, from a single original sample. Normal resampling is very similar to bootstrapping as it is a special case of the normal shift model—one of the assumptions for bootstrapping (Westfall et al., 1993). Both bootstrapping and normal resampling both assume that samples are drawn from an actual population (either a real one or a theoretical one). Another similarity is that both techniques use sampling with replacement.

Ideally, you would want to draw large, non-repeated, samples from a population to create a sampling distribution for a statistic. However, limited resources may prevent you from getting the ideal statistic. Resampling means that you can draw small samples repeatedly from the same population. As well as saving time and money, the samples can be quite good approximations for population parameters.

### 2. Permutation Resampling

Unlike bootstrapping, **permutation resampling** doesn't need a "population"; resampling is dependent only on the assignment of units to treatment groups. The fact that you're dealing with actual samples, instead of populations, is one reason why it's sometimes referred to as the *gold standard* bootstrapping technique (Strawderman and Mehr, 1990). Another important difference is that permutation resampling is a without replacement sampling technique.

### 3. Cross Validation

**Cross-validation** is a way to validate a predictive model. Subsets of the data are removed to be used as a validating set; the remaining data is used to form a training set, which is used to predict the validation set.

### References


Good, P. (2006). Resampling Methods: A Practical Guide to Data Analysis. Springer Science & Business Media.

Strawderman & Mehta (1990). On the validation of exact tests in software packages. Unpublished Manuscript.

Westfall, P. et al., (1993). Resampling-Based Multiple Testing: Examples and Methods for P-Value Adjustment. John Wiley & Sons.

## Bootstrap Sample

A **bootstrap sample** is a smaller sample that is "bootstrapped" from a larger sample. Bootstrapping is a type of resampling where large numbers of smaller samples of the same size are repeatedly drawn, with replacement, from a single original sample.

For example, let's say your sample was made up of ten numbers: 49, 34, 21, 18, 10, 8, 6, 5, 2, 1. You randomly draw three numbers 5, 1, and 49. You then replace those numbers into the sample and draw three numbers again. Repeat the process of drawing $x$ numbers $B$ times. Usually, original samples are much larger than this simple example, and $B$ can reach into the thousands. After many iterations, the bootstrap statistics are compiled into a bootstrap distribution. You're replacing your numbers back into the pot, so your resamples can have the same item repeated several times (e.g., 49 could appear a dozen times in a dozen resamples).

Bootstrapping is loosely based on the law of large numbers, which states that if you sample over and over again, your data should approximate the true population data. This works, perhaps surprisingly, even when you're using a single sample to generate the data.

- An *empirical bootstrap* sample is drawn from observations.
- A *parametric bootstrap* sample is drawn from a parameterized distribution (e.g., a normal distribution).

### Why Resample?

Ideally, you would want to draw large, non-repeated, samples from a population to create a sampling distribution for a statistic. However, you may be limited to one sample because of finances or time. This single sample method can serve as a mini population, from which repeated small samples are drawn with replacement over and over again. As well as saving time and money, bootstrapped samples can be quite good approximations for population parameters.

## Running the Procedure

Bootstrapping is usually performed with software (e.g., Stata or with the R Bootstrap package); The process generally follows three steps:

1. Resample a data set *x* times.
2. Find a summary statistic (called a bootstrap statistic) for each of the x samples.
3. Estimate the standard error for the bootstrap statistic using the standard deviation of the bootstrap distribution.

## Notation

- The number of bootstrap samples can be indicated with B (e.g., if you resample 10 times then B = 10)
- A bootstrap sample is identified by "star" notation: $x^*_1$, $x^*_2, \ldots, x^*_n$. This is similar to the notation for sample data, which is traditionally denoted by: $x_1, x_2, \ldots x_n$.
- A star next to a statistic, like $s^*$ or $\bar{x}^*$ indicates the statistic was calculated by resampling. A bootstrap statistic is sometimes denoted with a T, where $T^*_b$ would be the $B^{th}$ bootstrap sample statistic T.

## Bootstrap Percentile Method

The bootstrap percentile method is a way to calculate confidence intervals for bootstrapped samples.

With the simple method, a certain percentage (e.g., 5% or 10%) is trimmed from the lower and upper end of the sample statistic (e.g., the mean or standard deviation). Which number you trim depends on the confidence interval you're looking for. For example, a 90% confidence interval would generate a 100% – 90% = 10% trim (i.e., 5% from both ends). Or, put another (slightly more technical) way, you can get a 90% confidence interval by taking the lower bound 5% and upper bound 95% quantiles of the B replication $T_1, T_2, \ldots, T_B$.

A more complicated method is Efron's BCa method (see DiCiccio and Efron, 1993), which stands for Bias-Corrected and accelerated. As well as adjusting for bias, it also corrects skewness in the model. Other variants include Rubin's Bayesian extension and DiCiccio and Efron's ABC method.

This trimmed range for the statistic is the confidence interval for the population parameter of interest.

**References:**

DiCiccio, T.J. and Efron B. (1996) Bootstrap confidence intervals. Statistical Science, 11, 189-228.
Efron, B. and Tibshirani, R. (1993) An Introduction to the Bootstrap. Chapman and Hall, New York, London. Rubin, D (1981). The Bayesian bootstrap. Annals of Statistics 9 130–134.

## Markov Chain Monte Carlo

**Markov Chain Monte Carlo** is a way to sample from a complicated distribution. What is a "complicated distribution"? One that's difficult, or near impossible to find probabilities for. These types of problems are called "**high-dimensional**" problems. One example of a high-dimensional problem is calculating the relative importance of pages on the World Wide Web. Not only are their billions of pages, but they are in a constant state of change. Markov Chain Monte Carlo can solve these types of problems in a reasonable amount of time.

Markov Chain Monte Carlo isn't a specific technique: it's actually a large class of sampling algorithms that includes the Metropolis algorithm, Gibbs sampling, and Slice Sampling.

### WHAT IS "HIGH DIMENSIONAL"?

The key to understanding MCMC is to understand **high dimensional space**. It's an entire chapter in some computer science books, so I won't go into more than just a brief overview here.

*A hypercube.*

Basically, high-dimensional space is anything beyond the 2-D and 3-D world we're used to dealing with. High dimensional space doesn't behave the way we think it should. For example, a 4-dimensional cube, the hypercube, collapses in on itself as it rotates. Similarly, a high-dimensional unit-radius sphere has a volume of zero; Cubature (numerical integration) is needed to find these volumes. A high dimensional math problem has data that lies on high dimensional spaces. Therefore, the data doesn't "behave" the way we think it should.

High-dimensions are difficult to comprehend (it's said that even Einstein had trouble visualizing hypercubes, although he did perform the math behind four and five dimensions) and their space increases exponentially with each additional dimension–making it practically impossible to add any numerical meaning to the results. It follows then, that finding probabilities for high-dimensional data is extremely difficult. The usual probability rules don't apply, so new ones are needed, like MCMC.

## Markov Chains

Markov chains are used to model probabilities, where all the information needed to predict future events can be encoded in the current state. Basically, two colored die represent the current state (red or black) and a colored ball chosen from a bag gives the probability of transitioning to the next state. In real life, the number of states is likely to be much higher than just two! One researcher, Hopcraft, gives the concrete example of speech, where the last *k* syllables uttered by the speaker can be analyzed, giving the probability of what the next word might be. The speech is moving from one state to another in such a way that the next word depends only upon the current state (i.e., what's being said now). The number of possible states for the last *k*-syllables could reach into the millions, which would be impossible to analyze by hand (or even with a standard computer algorithm).

More technically, this information can be placed into a matrix and a vector (a column matrix) called probability vectors. Many, many, many iterations later and you have a collection of probability vectors…making up the Markov chains.

The purpose of the Markov Chain Monte Carlo is to sample a very large sample space, one that contains googols of data items. One example of such a sample space is the World Wide Web. Analyzing the web for important of pages is behind search engines like Google, and they use Markov chains as part of their analytics. Essentially, they send a bot out on (a literal) random walk of the web, following links wherever they go. These random walks are logged in a matrix, along with probabilities of reaching a certain page.

## Why Markov Chain Monte Carlo?

OK, so we have a complicated distribution — like the world wide web. MCMC is a way to efficiently sample from this complicated distribution. Simple random sampling or similar methods (like picking a name from a hat) won't work, because you won't be able to easily de ne your population because it's growing exponentially.

Even if you did find a way to define the pages on the current WWW, the millions of pieces of new content generated in one minute from now will have a zero probability of being chosen. From a statistical standpoint, this is unacceptable. Enter Markov Chain Monte Carlo.

The "Monte Carlo" part of MCMC is just a moniker for the type of sampling we're doing — which is sampling based on random walks. Monte Carlo is a gambling resort in Monaco, so it could just as easily have been called "Markov Chain Las Vegas" or just "Markov Chain Sampling."

## MCMC and Limiting Distributions

MCMC uses **random walks** (essentially a huge random number generator) to estimate probabilities with the same limiting distribution as the random variable you're trying to simulate. As time $n > \infty$, a Markov chain has a limiting distribution $\pi = (\pi_{j\,j})_{\in s}$. A value for $\pi$ can be found by solving a series of linear equations.

## References:


Blum, J. et. al. (2011). Foundations of Data Science. Retrieved August 11, 2017 from: https://www.cs.cornell.edu/jeh/bookMay2015.pdf

Agresti A. (1990) Categorical Data Analysis. John Wiley and Sons, New York.

Jerrum, M., & Sinclair, A. (1996). The Markov chain Monte Carlo method: an approach to approximate counting and integration. In D. S. Hochbaum (Ed.), Approximation algorithms for NP-hard problems (pp. 482–519). PWS Publishing. Kotz, S.; et al., eds. (2006), Encyclopedia of Statistical Sciences, Wiley.

# Acceptance-Rejection Sampling

**Acceptance-Rejection sampling** is a way to simulate random samples from an unknown (or difficult to sample from) distribution (called the **target distribution**) by using random samples from a similar, more convenient probability distribution. A random subset of the generated samples is rejected; the rest are accepted. The goal is for the accepted samples to be distributed as if they were from the target probability distribution.

Let's say you wanted to sample from the inverse of a particular function, but an explicit inverse doesn't exist. You could use a similar distribution—making sure that the two distributions look as similar as possible, and both functions have the same support (Carroll & Hamrick, 2011). In other words, both **cumulative distribution functions** (which give you the cumulative or additive probability associated with a function) vanish or don't vanish over the same set of real numbers)

Acceptance-rejection sampling is a Monte Carlo method that has largely been replaced by newer Markov Chain Monte Carlo methods. It's usually only used when it's challenging to sample individual distributions within a larger Markov chain (Christensen et al., 2011).

## General Steps

General steps for sampling from a probability distribution with cumulative distribution function f(x), using a second function g(x) (Dieter & Ahrens, 1974) are:

1. Generate a sample x from a distribution that has a density proportional to g(x), where $g(x) \geq f(x)$ for all x.
2. Generate a [0,1)-uniform random number u.
   If $u > f(x)/g(x)$ reject the sample and return to Step 1; Otherwise accept x as a sample.

Dieter and Ahrens note that although the procedure is relatively simple, some mathematical skill is needed to choose an appropriate function g(x).

**References**

Christensen, R. et al., (2011). Bayesian Ideas and Data Analysis: An Introduction for Scientists and Statisticians. CRC Press.

Dieter, U. & Ahrens, J. (1974). Acceptance-Rejection techniques for sampling from the gamma and beta distributions.

Office of Naval Research Technical Report No. 83. Retrieved December 9, 2017 from: https://statistics.stanford.edu/sites/default/les/CHE%20ONR%2083.pdf

Glasserman, P. (2004). Monte Carlo Methods in Financial Engineering. Springer Science & Business Media.

Carroll, R. & Hamrick, J. (2011). Wolfram Demonstrations Project. Acceptance/Rejection sampling. Retrieved December 9, 2017 from: http://demonstrations.wolfram.com/AcceptanceRejectionSampling/

# Importance Sampling

**Importance sampling** is a way to predict the probability of a rare event. Along with Markov Chain Monte Carlo, it is the primary simulation tool for generating models of hard-to-define probability distributions.

Rare events can usually be found on the tails of probability distributions. For example, on a bell curve for IQ, the Albert Einsteins of the world are found above three standard deviations from the mean. The rarity of finding results like this makes it extremely difficult to sample large enough numbers for any meaningful statistical analysis. In addition, the probability distribution for these rare events are going to look markedly different from the bell curve. Although predicting when another Einstein might be born is probably not that critical, predicting other rare events — like fatigue in engineering structures or landfall for category 5 hurricanes — can be a matter of life and death.

As well as finding probabilities in tails, importance sampling can also be used to find expectations of random functions.

## Biasing Density

One way to produce large enough samples is to change the probability density function to generate more rare events. This alternate density function is derived from the original function of interest (in the above example, the bell curve) and is usually called the **biasing density**. The end goal is to reduce the variance of your estimates. The basic steps are:

1. Choose a model for the process you want to study (i.e., derive the biasing density function and define the model's parameters (e.g., the mean and variance)),
2. Draw random samples from the parameterized model,
3. Run your statistical analysis on the biasing density function,
4. Modify those results to reflect the changes you made to the probability distribution.
5. Analyze the output.

## Importance Sampling and Monte Carlo Procedures

Importance sampling speeds up Monte Carlo procedures for rare events (a "**Monte Carlo procedure**" is sampling based on random walks). As it speeds up the process, it's sometimes referred to as "fast simulation using importance sampling." It's also called a "forced Monte Carlo procedure" because it's forcing the Monte Carlo procedure to behave somewhat abnormally.

If you're using Monte Carlo procedures, you're more than likely using software because of the large number of computations involved. Many statistical software packages include Monte Carlo algorithms, including Minitab, R and SPSS.

## Formulas

The formulas behind Importance Sampling are somewhat esoteric, mainly because of the calculus involved. As a (relatively) simple example, let's say you wanted to create an expectation for some function, f: $\mu_f = \mathcal{E}_p[f(X)]$, with

$$\mu_f = \int f(x)p(x)dx$$

Then for any probability density function q(x) that satisfies q(x) > 0 when $f(x)p(x) \neq 0$, you have:

$$\mu_f = \mathscr{E}_q[w(X)f(X)]$$

Where:

- $w(x) = p(x)/q(x)$
- $\mathscr{E}_q[x]$ = expectation with respect to q(x).

You can now use a sample of independent draws from q(x) to estimate $\mu_f$ by

$$\hat{\mu}_f = \frac{1}{m}\sum_{j=1}^{m} w(x^{(j)})\, f(x^{(j)}).$$

**References:**


Neal, R. M. (2001). Annealed Importance Sampling. Statistics and Computing (11) 125-139.

Oh, M.-S. and Berger, J. O. (1992). Adaptive Importance Sampling in Monte Carlo Integration. Journal of Statistical Computation and Simulation (41) 143-168.

Srinivasan, R. (2013). Importance Sampling: Applications in Communications and Detection. Springer Science & Business Media.

Tokdar, S. & Kass, R. (2009). Importance Sampling: A Review. Retrieved 8/18/2017 from: http://www2.stat.duke.edu/~st118/Publication/impsamp.pdf

# Latin Hypercube Sampling

| A | B | C |
|---|---|---|
| C | A | B |
| B | C | A |

**Latin Hypercube Sampling** (LHS) is a way of generating random samples of parameter values. It is widely used in Monte Carlo simulation, because it can drastically reduce the number of runs necessary to achieve a reasonably accurate result.

LHS is based on the **Latin square design**, which has a single sample in each row and column. A "hypercube" is a cube with more than three dimensions; the Latin square is extended to sample from multiple dimensions and multiple hyperplanes.

## The Method Behind Latin Hypercube Sampling

One-dimensional Latin hypercube sampling involves dividing your **cumulative density function (CDF)** into *n* equal partitions; and then choosing a random data point in each partition.

As a simple example, let's say you needed a random sample with 100 data points. First, divide the CDF into 100 equal intervals. If your distribution starts at 0 and ends with k, your first data point would be selected from the interval between (0,k/100). The second data point would be from the interval (k/100, 2k/100), your third from (2k/100, 3k/100), and so on. In each interval you would randomly select one point, giving you 100 different points. Two-dimensional Latin hypercube sampling is not much more complicated and is usually performed with software. Assuming your two variables, x and x are independent, you follow the one-dimensional method to come up with one dimensional samples for $x_1$ and $x_2$ separately. Once you have two lists of samples, you combine them, randomly, into two-dimensional pairs.

For n-dimensional Latin hypercube sampling the same method is used.

### WHY USE LATIN HYPERCUBE SAMPLING?

Latin Hypercube Sampling is typically used to save computer processing time when running Monte Carlo simulations. Studies have shown that a well-performed LHS can cut down on processing time by up to 50 percent (versus a standard Monte Carlo importance sampling).

LHS is therefore more important when working with slow operating systems and software than when doing analysis on faster devices. Some have gone so far as to suggest that the modern computers available to almost any researcher have made LHS obsolete, but it is still widely used. Although it does not make as big a difference in analysis as it did in the days of slow computer systems, it still leads to marginally more accurate results (in terms of true variability) given any amount of processing time.

## References

Olsson, A. et al. (2003). On Latin hypercube sampling for structural reliability analysis. Structural Safety: Volume 25,
Issue 1, January, Pages 47-68. Retrieved January 5, 2018 from:
https://doi.org/10.1016/S0167-4730(02)00039-5

Xin, L. (2014). Numerical Methods for Engineering Design and Optimization: Latin Hypercube Sampling (LHS)
Retrieved January 1, 2018 from:
https://users.ece.cmu.edu/~xinli/classes/cmu_18660/Lec25.pdf

www.ingramcontent.com/pod-product-compliance
Ingram Content Group UK Ltd.
Pitfield, Milton Keynes, MK11 3LW, UK
UKHW022017190726
13853UKWH00005B/1974